Textbooks in Electrical and Ele

CW01566990

Instrumentation for Engineers and Scientists

John Turner

Chief Research Scientist
Transport Research Laboratory

and

Martyn Hill

Mechanical Engineering Department
University of Southampton

OXFORD
UNIVERSITY PRESS

Preface

The science of instrumentation is of fundamental importance to engineers and scientists. Without measurement there can be no experimentation, and no control of even the simplest engineering system. Sensors are the eyes and ears of the technologist—even the sense of smell is used on occasion.

Without sensors and their associated signal conditioning, processing, data analysis, and storage systems we should live in a cold, dark, hostile world. There would be no transport other than that provided by animals, and anyone unlucky enough to fall ill would be offered only the most primitive medical treatment.

The progress that has been made in almost all areas of technology can be explained in terms of the rate at which the accompanying instrumentation has evolved. Of all engineering challenges, that of providing cheap, reliable, efficient measurement devices is often the hardest. In automotive engineering for example, microprocessors have proliferated for tasks such as engine management, largely because the cost of processing power (often expressed as the *price*-performance ratio or *PPR*) has fallen so dramatically. In the last decade the PPR for microprocessors has improved by a factor of over 1000. Similarly, electric motors have proliferated on cars: many tasks such as seat and mirror adjustment, and opening windows etc., are now motorized. Again the reason is linked to the PPR: for electric motors it has improved by between 50 and 100 depending on the motor type.

Sensors have not become cheaper by anything like the same amount. Most studies reveal that the sensor PPR has only improved by around a factor of 10 in the last 20 years. This explains why so many driving tasks, such as speed control, turning on and off windscreen wipers, headlights, heated windows and so on are usually controlled manually.

However, things are about to change. The silicon fabrication techniques that have been developed for manufacturing microprocessors are now being exploited for sensor manufacture, and the resulting devices (known as micromachined sensors) are much cheaper and more robust than transducers manufactured using conventional techniques. Since micromachined sensors are fabricated using the methods developed for circuit fabrication, they can incorporate signal-processing systems within the sensor package. Amplifiers, filters, and analog-to-digital converters can all be included, as well as subsidiary sensors: for example, to measure the transducer temperature and correct its output for thermal drift before communicating it to the outside world.

The advent of micromachined measurement devices is likely to bring about large changes in sensor PPR, and consequently large changes in the way in which measurement systems are incorporated into mechanical products such as cars, domestic appliances and laboratory equipment. The science of instrumentation, already of such fundamental importance, is likely to become even more significant as greater and greater intelligence is built into products. This vital topic is however not often taught at undergraduate level to non-electrical engineers, and herein lies the genesis of this book. Both of the authors have for some time been involved with teaching and supervising final-year mechanical engineering undergraduates. In our opinion it is vital that engineers and scientists have a good working knowledge of the principles involved in measurement and analysis. If the best performance is to be obtained from a measurement or control system, its weaknesses must be thoroughly understood as well as its strong points. It is a truism, but one worth repeating, that the most critical features of any electronic device or system are the ones the manufacturer has chosen NOT to include on the data sheet! Our aim has been to produce a university-level text which is accessible to final year students from general engineering or science backgrounds: and also to produce a practical guide to instrumentation which will be of use to practising engineers and scientists.

Southampton   J. D. T. and M. H.
November 1998

Contents

3 Displacement sensing

4 Velocity and acceleration transducers

5   Strain measurement techniques

9   Signal conditioning circuits

General design of instrumentation systems—error analysis

1.1 Introduction

Instrumentation is a subject of fundamental importance to engineering, science, and medicine. From the student undertaking a laboratory investigation to the operators of a nuclear power plant, accurate measurements are an essential prerequisite to the understanding and control of all physical processes.

In general, an instrumentation system may fall into one of two categories. First, there are laboratory or experimental measurement techniques used for research and development. The most important consideration faced by the designer of a measurement system intended for experimental work is its performance. In acquiring research data a high degree of repeatability, accuracy, linearity and reliability are required. The cost of this kind of system is usually (but not always) of secondary importance.

The second sort of measurement system is that which forms part of a well-understood device, usually a commercial product. Examples of this kind of instrument can be seen on a motor vehicle, where the driver is provided with a speedometer to help control the vehicle, a petrol gauge to indicate when fuel is required, and a milometer or other indicator to show when maintenance is needed.

For a well-understood system such as a motor vehicle, a lower degree of instrument performance than that required for research is usually sufficient. For example, the average automotive speedometer probably has an accuracy of no more than 10%. This low resolution is entirely adequate to control the vehicle. In general, measurement systems supplied as part of a product are of lower performance than those used for experimental work. The main reason for this is so that the complete system can be produced at a reasonable cost.

In the automotive example three measurement system types were identified:

(1) parameters used for control, such as speed,
(2) warning devices, such as a fuel gauge, and
(3) condition monitoring data, such as that provided by a milometer or run-time indicator.

At least one of these categories of information will be present in all measurement systems forming part of a product.

Measurements are carried out by means of transducers. A transducer is a device which converts energy from one form to another. *Output transducers*, or *actuators*, convert electrical, pneumatic, hydraulic, or other forms of energy into mechanical force or displacement. *Input transducers*, or *sensors*, convert state parameters such as temperature, pressure, force, magnetic field strength etc. into (usually) electrical energy, since this is generally the most convenient form for measurement or signal processing. Since the subject of this book is instrumentation, we shall mainly be concerned with sensors rather than actuators.

Once information regarding the variation of a physical parameter has been produced in an electrical form by the action of a sensor, the instrumentation designer has then to consider a number of subsidiary problems. The output of most sensors takes the form of a small electrical signal. This signal is most often a varying voltage, but can also be a varying current or charge. The voltage form is usually the most convenient for subsequent analysis. Thus, signal conditioning circuits may have to be constructed to convert the signal to voltage form, amplify it, and apply filtering if required to remove unwanted noise.

At this stage information about the behaviour of the parameter of interest is available to the engineer in the form of an analogue voltage. A decision must then be made regarding the best way to proceed. In the past, most signal analysis was carried out by means of analogue circuits. For example, until the mid-1970s it was common to examine the spectrum of a signal by means of analogue systems, such as a *swept-filter analyser*. Analogue techniques are still sometimes used, but the widespread availability of personal computers (PCs) running powerful analysis software (such as MATLAB) has meant that signals are normally converted to digital form if anything more than rudimentary signal processing is required.

The instrumentation designer has to decide therefore whether the data should remain in analogue form, or whether the required signal processing and analysis are better carried out in a digital format. There is no doubt that the digital approach greatly facilitates signal analysis, but the process of digitization is not always straightforward. Care has to be taken over *aliasing, sample rate*, and the choice of *binary word size* (which determines *resolution*). These questions are discussed in detail in a later chapter.

The kind of signal processing ultimately carried out will depend upon the information required from the signal. Frequently all that is needed is the amplitude of the signal, to indicate the amplitude of the force, pressure etc. being measured. In such cases a straightforward display of the instantaneous signal, or its root-mean-square (rms) or peak value, is all that is necessary. However, it may be necessary to use signal processing techniques such as smoothing, averaging, correlation or the Fast Fourier Transform (FFT) to extract the required information.

1.2 Generalized instrumentation design

An instrument can be defined as a system which maintains a prescribed relationship between the parameter being measured and some other physical variable. The second variable is used as a means of communicating information regarding the first, either to a human observer or to some other measurement or control system. An indication of how well a prescribed functional relationship is maintained can be had from the static and dynamic calibration of the instrument concerned.

A measurement system may be broken down into its functional elements as shown in Fig. 1.1. Every instrumentation system contains some or all of these functional blocks. If the behaviour of the elements is known an assessment of the performance of the complete system can be made.

Information regarding the state of a physical system is obtained through a change in one of the properties of the system. For example, in a vibrating system changes in the amplitude, frequency or phase of vibration all convey information about the state of the system. The physical parameter being measured, which forms the input to the generalized measurement system of Fig. 1.1, is known as the *measurand*. The primary sensing element in any measuring system is that which first receives energy from the object being measured, and which produces an output according to a well-understood relationship with the measurand.

A measurand is almost always disturbed by the act of measurement, although in a well-designed instrument this effect is minimized. For example, the usual way of instrumenting a vibrating system is by means of one or more accelerometers. However, most accelerometers have a mass of 100 g or so, and one of the effects of adding mass to a vibrating system is to

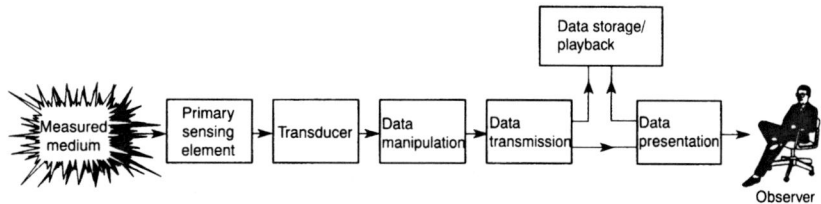

Fig. 1.1 The functional elements of an instrumentation system.

alter its dynamic characteristics. The extent of the alteration depends on the relative masses of the vibrating system and the sensors.

At the other end of the scale, the atomic structure of a crystalline material such as silicon may be investigated by means of an electron microscope. Unfortunately, the high-energy electrons used to form an image of the material displace atoms from the crystal being studied. (At the level of quantum effects, this is a consequence of Heisenberg's Uncertainty Principle). Once again, we see that the act of measurement causes changes in the system under investigation.

The change induced in the primary sensing element by the action of the measurand is communicated to a transducer, or device which converts energy from one form to another. The output of most transducers is electrical, since electronic signals are most conveniently processed or transmitted. Transducers may be grouped into three categories, as shown by Fig. 1.2. In the first type, the same form of energy exists at both the input and the output. This kind of transducer is known as a *modifier*, since energy is modified rather than converted to a different form. An example might be a bandpass filter, which can be used to measure the energy within a particular bandwidth in an electrical signal.

The second category of transducers are those known as *self-generating* or *self-exciting*. In these devices electrical signals are produced directly from a non-electrical input, without the application of any external energy. Examples of self-generating sensors include thermocouples, photovoltaic cells, and devices which rely on the piezoelectric effect, such as some types of accelerometer. Self-generating transducers are usually characterized by their very low output energy. They normally require a number of amplification stages to increase the signal amplitude to a useful level.

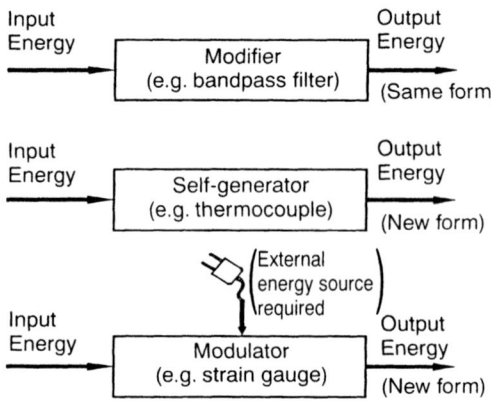

Fig. 1.2 Three classes of transducer.

The third type of transducer also produces an electrical output from a non-electrical input, but requires an external source of energy to function. Members of this group are known as *modulating* transducers. The strain gauge is a familiar example, in which a mechanical deformation is used to control a variable resistance, and a signal voltage is generated by passing a current through the device.

1.3 Error analysis and the performance of a measurement system

The performance of any instrument depends upon both its static and dynamic characteristics. In the case of rapidly varying quantities, the relationship between the input and the output of an instrument is usually expressed by means of differential equations. However, for many applications the parameters being measured vary sufficiently slowly for the dynamic effects to be neglected.

The errors caused by nonlinearity, drift, resolution errors, and repeatability are usually considered as static characteristics. Hysteresis, settling time, and variations in the response at different frequencies are dynamic effects. The total system response is obtained by combining the static and dynamic responses.

All instruments suffer from inherent inaccuracies. Every measurement has an associated error (except for the counting of discrete events, which may have no associated error). To estimate the uncertainty in a measurement, it is necessary to know the form of the error. In general, any error is a combination of two: a systematic error, in which all of the readings are biased, and a random error, through which repeated readings are found to be scattered around a mean representing the true value.

Systematic errors usually arise from an unsatisfactory experimental method or measurement system design. For example, it is common to find that an instrument does not quite return to zero when the parameter being measured assumes a zero value. This effect is known as a *zero error*, and will cause a systematic shift—or systematic error—in any reading made using that instrument.

Random errors arise from random or chance causes, and must be treated using statistical methods. For example, suppose the mass of an object is measured using a sensitive balance. After weighing the object several times, the readings will be found to form a group around a mean value. This may, for example, be because the currents of air in the room containing the balance have disturbed the mechanism during weighing. To specify the precision of a mean value obtained from a group of readings, some indication of the scatter of the readings around the mean is required. To indicate the size of a random error the *standard deviation* about the mean is specified.

1.3.1 Random errors

In many physical processes the distribution of errors about a true value will be Gaussian as shown in Fig. 1.3. Generally speaking, at least 25 readings are needed before a Gaussian (or normal) distribution may be assumed. The value of the Gaussian distribution lies in the fact that, if a large enough number of readings has been taken, 68% of the readings will lie within one standard deviation of the mean, and 95% within two standard deviations. In general, if n measurements $x_1, x_2, \ldots, x_n$ of a physical parameter are taken under the same conditions, then the

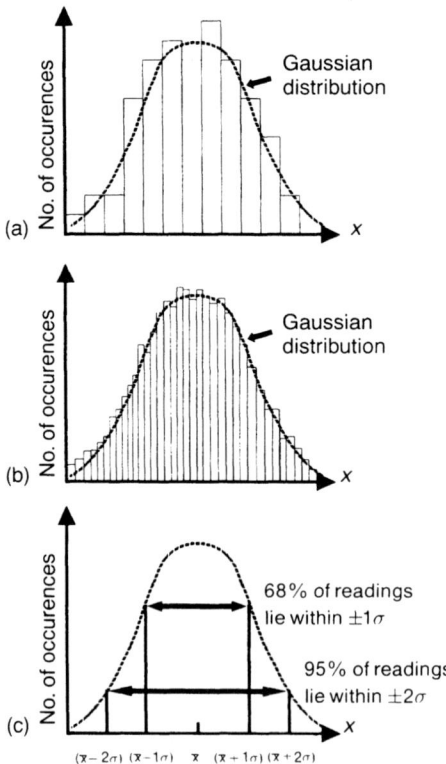

Fig. 1.3 Random data and the Gaussian (Normal) distribution. (a) If >25 readings are plotted as a histogram an approx. Gaussian distribution results. (b) More readings gives a better approximation. (c) Confidence limits.

best estimate of the value of the parameter under those conditions is the mean value $\bar{x}$, which is given by:

$$\bar{x} = \frac{\sum_{i=1}^{n} x_i}{n}. \tag{1.1}$$

The standard deviation σ of the set of readings (referred to as a sample by statisticians) is given by:

$$\sigma = \sqrt{\left(\frac{\sum_{i=1}^{n} d_i^2}{n}\right)} \tag{1.2}$$

$$d_n = x_n - \bar{x}, \tag{1.3}$$

where the values $d_1 \ldots d_n$ are defined as the difference between the individual readings and the mean.

For a random process (which has a Gaussian distribution), the bounds $\pm 1\sigma$ and $\pm 2\sigma$ show how far an individual reading is likely to be from the true value. The normal procedure is to take several readings and find their mean. How far this *sample mean* is from the true value is of much more concern than the uncertainty associated with any individual reading. It should be emphasized that there is no way of knowing what the true value is, since an infinite number of readings would be required to calculate it. Obviously, the extent to which the sample mean departs from the true value depends not only on the spread of the individual values (the standard deviation), but also on the number of readings taken. However, it is possible to specify the probability that the sample mean lies within a specific range of the true value. This range is known as the *standard error*, s_m, and is calculated from:

$$s_m = \frac{\sigma}{\sqrt{n-1}}. \tag{1.4}$$

The probability that the mean of a given sample from a Gaussian population lies within $\pm s_m$ of the true value is 68%. The probability that the mean lies within $\pm 2s_m$ of the true value is 95%. The standard error s_m therefore an estimate of how close the mean of a sample set of values x is to the true value.

1.3.2 Systematic errors

From eqn (1.4) it can be seen that by taking a large enough number of readings the random error on a measurement may be made as small as desired. However, when a systematic error occurs all the measurements are systematically shifted in one direction, and the process of taking a number of readings and finding their mean value will not improve the

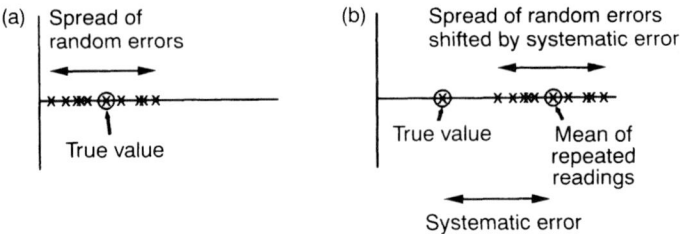

Fig. 1.4 The effects of random and systematic errors. (a) Random errors. (b) Combined random & systematic errors.

accuracy of the measurement. Figure 1.4 shows the spread of readings caused by a random error (i.e. the *precision*), and also how the randomly-distributed readings may be shifted by a systematic error (the *bias*), so that the mean value is itself in error. If we know the characteristics of the process under investigation, we can put bounds on the error of the single measurement, although we cannot tell what the error itself is (since this would imply that we knew the true value). Statements about the accuracy of a particular measurement can only be made in terms of the *precision* and *bias* of the measurement process.

Earlier in this chapter a zero offset was given as an example of a systematic error. A further example is the case of a clock which runs too slowly or too fast. It is not feasible to give a complete list of all possible systematic errors, since each instrumentation system is liable to its own particular hazards. Systematic errors are cumulative, so that if for example a measurement A is an additive function of the quantities x, y, and z, i.e. $A = f(x + y + z)$, then the *maximum possible* value of the systematic error is $(\delta x + \delta y + \delta z)$, where δx, δy and δz are the errors in x, y, and z. However, this is usually rather a pessimistic approach, since it requires that all the systematic errors operate in the same direction. It is much more likely that, if the errors are independent, some of them will systematically increase the reading and others will systematically decrease it. It is therefore more usual to quote the *probable systematic error* ΔA, which is evaluated from

$$\Delta A = \sqrt{\delta x^2 + \delta y^2 + \delta z^2}. \tag{1.5}$$

1.3.3 Indirect measurements

It is often necessary to estimate the value of a quantity from measurements of other parameters which have a reasonably well-defined functional relationship with the desired quantity. If A_m denotes a quantity which is estimated from other measurements a_1, a_2, ... by the use of an expression of the form:

$$A_m = f(a_1, a_2, ...)$$

then the estimate of A_m is:

$$\hat{A}_m = f(\hat{a}_1, \hat{a}_2, ...),$$

where $\hat{a}_1$, $\hat{a}_2$, ... are the estimated values of the means of a_1, a_2, etc.

If the measurements a_1, a_2, etc. are subject to random error processes, the associated *variances* (variance = standard deviation²) will be $\sigma^2(a_1)$, $\sigma^2(a_2)$, etc. The variance associated with the indirectly measured quantity A_m is given by:

$$\sigma^2(A_m) = \left(\frac{\partial A_m}{\partial a_1}\right)^2 \sigma^2(a_1) + \left(\frac{\partial A_m}{\partial a_2}\right)^2 \sigma^2(a_2) + \cdots. \tag{1.6}$$

In eqn (1.6) all variables are assumed to be independent, and higher order terms have been neglected.

In some cases an indirect measurement has to be based on quantities which suffer from systematic uncertainties. If $\Delta(A_m)_1$ is the component of systematic uncertainty in A_m due to the systematic error Δa_1 in a_1, then:

$$\Delta(A_m)_1 = \left|\frac{\partial A_m}{\partial a_1}\right| \Delta a_1.$$

There is no rigorous method for combining components of systematic uncertainty to give the total systematic uncertainty $\Delta(A_m)$. Two approaches are used; in one the components are simply added to estimate the maximum possible uncertainty:

$$\Delta(A_m) = \left|\frac{\partial A_m}{\partial a_1}\right| \Delta a_1 + \left|\frac{\partial A_m}{\partial a_2}\right| \Delta a_2 + \cdots. \tag{1.7}$$

This approach is likely to result in an overestimate.

The alternative is to combine the systematic uncertainties in quadrature:

$$\Delta(A_m) = \sqrt{\left[\left(\frac{\partial A_m}{\partial a_1}\right)^2 (\Delta a_1)^2 + \left(\frac{\partial A_m}{\partial a_2}\right)^2 (\Delta a_2)^2 + \cdots\right]}. \tag{1.8}$$

This method tends to underestimate the uncertainty. The best approach is probably to define upper and lower uncertainty bounds by applying both eqns (1.7) and (1.8). It should not be forgotten that it is also possible for an uncertainty to exist regarding the nature of the relationship between A_m and the directly-measured quantities a_1, a_2, etc. If this is the case it may be partially dealt with by the introduction of an *uncertainty factor* into the earlier equations.

1.3.4 Combining errors on sums, differences, products, and exponentials

The preceding section dealt with the subject of indirect measurements, and led to eqns (1.7) and (1.8) which define the upper and lower uncertainty bounds associated with an indirect measurement.

A somewhat less rigorous approach is adequate in many circumstances. If a quantity A is evaluated from measurements x, y, z, and the associated uncertainties are δx, δy, and δz, eqn (1.5) gives an estimate of the probable error. If the quantity A is calculated from the expression:

$$A = x + y,$$

the probable error ΔA is:

$$\Delta A = \sqrt{\delta x^2 + \delta y^2},$$

where x and y are measured quantities with the associated errors δx and δy.

If the quantity A is a difference, i.e. $A = x - y$, the error is still found from eqn (1.5). Although y is negative, the error δy is as likely to be positive as negative. Thus the resultant error is again found from the square root of the sum of the errors squared:

$$\Delta A = \sqrt{\delta x^2 + \delta y^2}.$$

If A is a product of the measured quantities, i.e. $A = xy$, a slightly different procedure must be followed. Consider the effect of increasing x to $x + \delta x$. Then:

$$A + \Delta A = (x + \delta x)y$$

$$= xy + \delta xy.$$

That is, $\Delta A = \delta xy$. If both sides are divided by A we have:

$$\frac{\Delta A}{A} = \frac{\delta xy}{A}$$

Therefore

$$\frac{\Delta A}{A} = \frac{\delta xy}{xy}$$

$$= \frac{\delta x}{x}.$$

Thus, a fractional increase in either x or y produces the same fractional increase in A. The error ΔA is once again evaluated from eqn (1.5), but this time fractional errors must be used. We therefore have:

$$\frac{\Delta A}{A} = \sqrt{\left(\frac{\delta x}{x}\right)^2 + \left(\frac{\delta y}{y}\right)^2}.$$

If there are more than two factors involved the same rule applies, and the error in A is calculated from the square root of the sum of the squares of the fractional errors. Expressed mathematically, if

$$A = x_1 x_2 x_3 x_4 \ldots x_n$$

then:

$$\frac{\Delta A}{A} = \sqrt{\sum_{i=1}^{n}\left(\frac{\delta x_i}{x_i}\right)^2}. \tag{1.9}$$

If A is a ratio, i.e. $A = x/y$, the procedure to be followed can once again be deduced by considering the effect of increasing x to $x + \delta x$:

$$A + \Delta A = \frac{x + \delta x}{y}$$

$$\therefore A + \Delta A = \frac{x}{y} + \frac{\delta x}{y}$$

$$\therefore \Delta A = \frac{\delta x}{y}$$

$$\therefore \frac{\Delta A}{A} = \frac{\delta x}{y} \cdot \frac{y}{x}$$

$$\text{i.e.} \quad \frac{\Delta A}{A} = \frac{\delta x}{x}.$$

Once again we see that the effect of a fractional change in x or y is to produce the same fractional change in A. Equation (1.9) is therefore used again to evaluate the error in A.

1.4 Conclusions

The principles outlined in this chapter are applicable to the full range of measurement practice. The following chapters describe measurement techniques for a range of important engineering parameters.

2.1 Introduction

Temperature is an important parameter for engineers, and is very apparent to our senses. The *concept* of temperature is elusive, however, and can cause difficulty. It is less easy to define a meaningful scale for temperature than it is for, say, length. This is probably because it is less easy to relate temperature to the fundamental dimensions of mass, length, and time than it is for most other common measurements, such as pressure, flowrate, strain, etc. Temperature cannot be measured directly, but has to be detected by observing the effect it produces, such as the expansion of a liquid, and assumptions have to be made about the linearity of the effect observed.

The first step towards the definition of a useful temperature scale came with the recognition by Galileo of heat as a measurable quantity, which flows from a hot to a cold body. Galileo's *thermoscope* was probably the first practical thermometer.

Most electronic temperature sensors used in engineering fall into one of two classes: resistive or thermoelectric. Resistive devices may be either metallic or semiconductor, and require some form of bridge circuit for signal conditioning since they are modulating transducers. Thermoelectric sensors or thermocouples are self-generating, but their very low output means that an amplifier is always needed in practice. In the past bimetallic temperature sensors were very common, and these are still sometimes used. The thermal expansion of a solid can also be used as a temperature transducer, for example to control thermostat opening in water-cooled internal combustion engines. Infra-red emission or pyrometry can be used for temperature sensing, and is particularly useful for monitoring inaccessible or rotating components. A further temperature-based sensor is the *heat flux gauge*. These measure the rate at which heat is being transferred to or from a body, rather than its temperature.

The temperature of any system is a measure of its energy. This is best described by the Principle of Equipartition, which states that for a system in thermal equilibrium with its surroundings, the mean kinetic energy per degree of freedom of the particles which go to make up the system is:

$$\text{mean kinetic energy} = \frac{kT}{2}.$$
(2.1)

In this equation k is Boltzmann's constant and T is the absolute temperature in degrees kelvin (K).

Equation (2.1) gives rise to Brownian motion (in mechanical systems), Johnson noise (in electronic systems), and radiation noise (in thermal systems).

The most direct realization of a scale of temperature using this approach is based on the ideal gas equation, which can itself be derived directly from the Principle of Equipartition by means of the Kinetic Theory of Gases [1]. The gas equation:

$$pv = RT \tag{2.2}$$

gives the pressure p of a specific volume[1] v of gas at a temperature T, with a constant of proportionality R which is equal to the number of molecules in the specific volume divided by Boltzmann's constant k. The constant volume gas thermometer, in which the pressure of a fixed mass of an inert gas is measured as a function of temperature, uses this approach. It forms the basis of most temperature measurement schemes. The only fixed point used is the triple point of water, defined as 273.16 K or 0.01 °C by international agreement.

Absolute measurements with gas thermometers are infrequently made, and are usually only undertaken by standards laboratories. For practical purposes a set of fixed points have been agreed which form the basis of the International Practical Scale of Temperature. Table 2.1 shows some of the fixed points, which are chosen for their ease of replication. Interpolation between them can be achieved by a temperature transducer once it has been calibrated with respect to the fixed points.

Table 2.1
Reference points used to define temperature scale

Temperature (°C)	Reference point	Measurement technique used
−182.97	Boiling point (BP) of oxygen	
0.01	Triple point of water	PRT
100.0	BP of water	
444.6	BP of sulphur	
960.8	Melting point (MP) of silver	Platinum thermocouple
1063.0	MP of gold	

[1] Specific volume $v = (\text{density})^{-1}$.

2.2 Resistive temperature transducers

Resistive temperature sensors are probably the most common type in use. They may be based on metals or semiconductors. The semiconductor versions are the most common and are probably the cheapest. They are sometimes known as *resistance temperature detectors* or RTDs. Metallic resistive temperature sensors offer better performance than semiconductor RTDs and may be preferred if high accuracy is required.

2.2.1 Metallic resistive temperature sensors

These transducers are similar in appearance to wire-wound resistors, and often take the form of a non-inductively wound coil of a suitable metal wire such as platinum, copper, or nickel. They may be encapsulated within a glass rod to form a temperature probe, which can be very small in size. An alternative form is shown in Fig. 2.1, in which a rectangular matrix of platinum is deposited on a ceramic substrate. Connections within the matrix are laser-trimmed to give a precise value of nominal resistance.

The variation of resistance R with temperature T for most metallic materials can be represented by an equation of the form:

$$R = R_0(1 + a_1 T + a_2 T^2 + \cdots + a_n T^n), \qquad (2.3)$$

where R_0 is the resistance at temperature $T=0$. The number of terms necessary in the summation depends on the material, the accuracy required, and the temperature range to be covered. Platinum, nickel, and copper are the most commonly used metals, and they generally require a summation containing at least a_1 and a_2 for accurate representation.

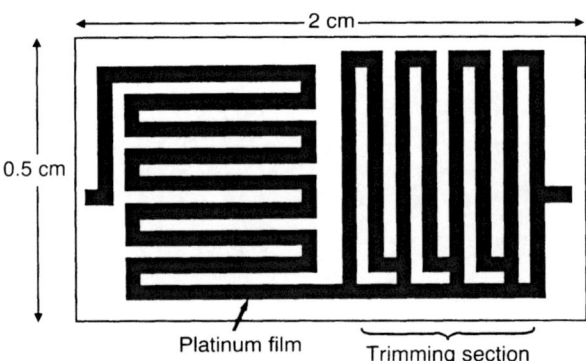

Fig. 2.1 Miniature platinum resistance transducer.

Tungsten and nickel alloys are also frequently used. In engineering applications it is often possible to model a metallic RTD using only constant a_1. With platinum for example, $a_1 \approx 0.004$ (R in Ω and T in K). If this value is substituted into eqn (2.3), and a_2, a_3, etc. are set to zero, the resulting nonlinearity is only around 0.5% over the temperature range -40 to $+140\,°C$.

The nominal resistance (R_0) of a metallic RTD can vary from a few ohms to several kilohms. However, $100\,\Omega$ is a fairly standard value. The resistance change of a metallic RTD can be quite large, and is typically up to 20% of the nominal resistance over the design temperature range.

2.2.2 Thermistors

Thermistors are small semiconducting transducers, usually manufactured in the shape of beads, disc, or rods. They are made by combining two or more metal oxides. If oxides of cobalt, copper, iron, magnesium, manganese, nickel, tin, titanium, vanadium, or zinc are used, the resulting semiconductor is said to have a *negative temperature coefficient (NTC) of resistane*. This means that as the temperature rises, the electrical resistance of the device falls. Most of the thermistors used in engineering are of this type, and they can exhibit large resistance variations. Typical values are $10\,k\Omega$ at $0\,°C$ and $200\,\Omega$ at $100\,°C$. This very high sensitivity allows quite small temperature changes to be detected. However, the accuracy of a thermistor is not as good as that of a metallic RTD, due to unavoidable variations in the composition of the semiconductor which occur during manufacture. Most thermistors are manufactured and sold with tolerances of 10 or 20%. Any circuit using semiconductor thermistors must therefore include some arrangement for adjusting out errors.

Thermistors are markedly nonlinear (unlike metallic RTDs). The resistance–temperature relation is usually of the form:

$$R = R_0 e^{\beta(1/T - 1/T_0)}, \tag{2.4}$$

where R is the resistance (in ohms) at temperature T (in Kelvin), R_0 is the nominal resistance at temperature T_0, and β is a constant which is characteristic of the thermistor material. The reference temperature T_0 is usually taken to be 298 K (25 °C), and β is of the order of 4000.

If eqn (2.4) is differentiated and rearranged, the temperature coefficient of resistance α at temperature T_0 can be found:

$$\alpha = \frac{(dR/dT)}{R} = -\beta/T_0^2, \tag{2.5}$$

where β is the characteristic temperature constant for the material and T_0 is in degrees kelvin. The units of the temperature coefficient of resistance α are ohms/degree kelvin (Ω/K). If β is taken as 4000, a typical value for α for a thermistor at 298 K (25 °C) is $-0.045 \, \Omega$/K.

Thermistors can be used within the temperature range from -60 to $+150$ °C. The accuracy can be as high as $\pm 0.1\%$. The main problem associated with thermistors is their nonlinearity as expressed by eqn (2.4).

Positive temperature coefficient (PTC) thermistors can also be made, using compounds of barium, lead, or strontium. PTC thermistors are usually only used to provide thermal protection for wound equipment such as transformers and motors. The characteristics of a PTC device have the form shown in Fig. 2.2. It can be seen that the resistance of a PTC thermistor is low (and reasonably constant) below the switching temperature T_R. Above this point the resistance rises spectacularly. In use, PTC thermistors are often embedded in the windings of the equipment to be protected, and are connected in series with the power supply. If the temperature becomes too high the resistance rises and power is effectively disconnected from the load. PTC thermistors are better than NTC types for this sort of thermal protection task since they are fail-safe: if a connection to a PTC sensor fails, the resulting high impedance will disconnect the power. If the same happens to a thermal protection circuit containing an NTC thermistor, a false 'low temperature' indication will be given, and full power is applied.

'Conventional' thermistors are manufactured as discrete components. However, it is also possible to print thermistors on to a suitable substrate using *thick film* fabrication techniques. Thick film thermistors are very low-cost and physically small, and have the further advantage of being more intimately bonded to the substrate than a discrete component.

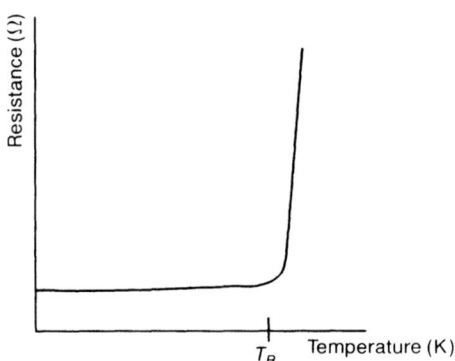

Fig. 2.2 Resistance temperature characteristic for PTC thermistor.

It has been shown [2] that thick film thermistors can have as good, if not better, performance than a comparable discrete component.

2.2.3 Resistance temperature sensor bridge circuits

Thermistors are modulating transducers and are normally used in a Wheatstone bridge circuit, which must include some form of bridge balancing arrangement as shown in Fig. 2.3. While the resistance changes exhibited by a metallic RTD are reasonably linear, those shown by semiconductor thermistors are markedly nonlinear. In both cases the resistance changes are large. Even if the sensor output is linear, the out-of-balance voltage measured using a bridge circuit is not necessarily linear for large changes in sensor resistance. Take for example the case of a $500\,\Omega$ platinum resistance thermometer which exhibits a $100\,\Omega$ resistance change over its design temperature range. If the sensor is included in a bridge with four equal arms, the out-of-balance voltage will be very nonlinear as a function of temperature. However, the fixed resistors R_1 and R_2 in Fig. 2.3 are of considerably higher resistance (about $\times 10$ is normal) than the platinum sensing resistors R_3 and R_4, and if care is taken to balance the bridge at the middle of its design temperature range rather than at one end, reasonable linearity may be achieved.

Resistance thermometer bridges may be excited with either AC or DC voltages. The current through the sensor is usually in the range from 1 to 25 mA. This current causes I^2R heating to take place, which raises the temperature of the thermometer above that of its surroundings and causes a so-called *self-heating error* to occur. The magnitude of this error depends on the heat transfer conditions (for instance, the conductivity of the surface to which the sensor is attached, and the presence or otherwise of a fluid flow).

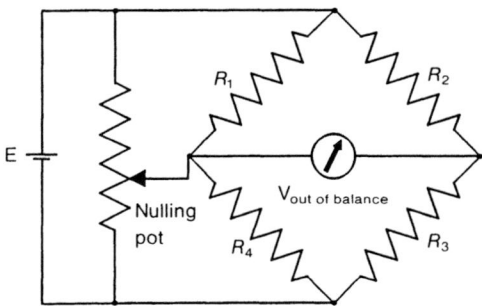

Fig. 2.3 Bridge circuit with balancing adjustment.

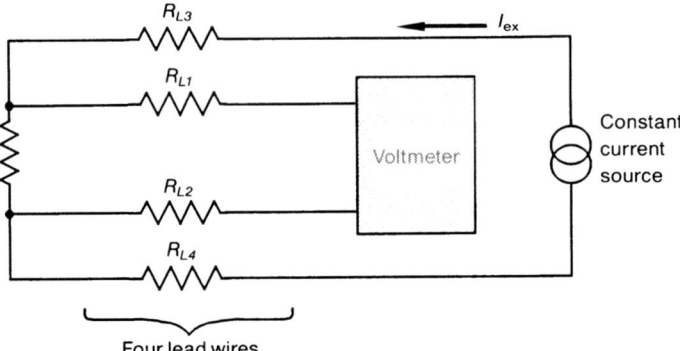

Fig. 2.4 Four-wire ohmmeter circuit.

An alternative to the classical bridge techniques for conditioning RTD sensors is the four-wire '*ohmmeter*' technique shown in Fig. 2.4. This is widely used with digital data acquisition systems, where any sensor non-linearity is corrected in the computer software. A precision current source is used, so resistance changes in the two connecting wires *L3* and *L4* have no effect on the sensor current I_{ex}. A high-impedance voltmeter is used, typically with a FET input of $> 200\,M\Omega$. This ensures that the currents through *L1* and *L2* are negligible, as are the lead-wire resistance errors.

2.3 Thermocouples

A thermocouple is a self-generating transducer comprising two or more junctions between dissimilar metals. The conventional arrangement is shown in Fig. 2.5(a), and it will be noted that one junction (the *cold junction*) has to be maintained at a known reference temperature, for instance by surrounding it with melting ice. The other junction is attached to the object to be measured.

Thermocouple materials are broadly divided into two arbitrary groups based upon cost. The groups are known as the *base metal* and *precious metal* thermocouples. The most commonly used industrial thermocouples are specified by type letters as shown in Table 2.2.

The arrangement of Fig. 2.5(a) is inconvenient because of the layout of leads and the need for a reference temperature. A more practical scheme is shown in Fig. 2.5(b). The two wires are laid out side by side, and are connected to a voltage measuring circuit. The junctions between the two wires and the voltmeter do not cause any error signal to appear so long as they are at the same temperature. Since there is no proper reference junction with this approach, the system is liable to give an erroneous

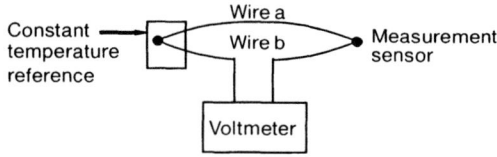

(a) Standard thermocouple arrangement.

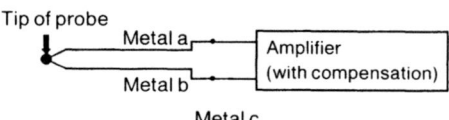

(b) Thermocouple arrangement with cold junction compensation.

Fig. 2.5 Thermocouple circuits.

output if the temperature of the surrounding environment changes. This is avoided by the use of so-called *cold junction compensation* (see later), in which the characteristics of the signal conditioning amplifier are modified by including a thermistor in the circuit.

The arrangement shown in Fig. 2.5(b) is usually applied whenever thermocouples are used. The two wires are often enclosed within a tube or flexible sleeve of stainless steel or copper for protection, although this increases the time constant of the system.

The main advantages of thermocouples are their wide temperature range, nominally from -180 to $+1200\,°C$ for a Chromel/Alumel device, and their linearity. Table 2.2 gives the characteristics of some of the most common commercially-available devices.

If a short section of tubing is made from butt-welded thermocouple materials and inserted into a pipeline, the temperature of a fluid flowing inside the pipe may be measured non-intrusively. Ready-made sensors of this type are available commercially [3] and are often used for experimental work.

2.3.1 Thermocouple compensation

As noted earlier, it is not normally practical to have thermocouple cold junctions maintained at a controlled reference temperature. However, with the cold junctions at ambient temperature, which may change, some form of cold junction compensation is required. Consider the arrangement shown in Fig. 2.6, which shows a thermocouple with its measuring junction at $t\,°C$ and its cold junction at ambient temperature. The thermocouple output is $E_{(a-t)}$, but what is required is the output that would be obtained if the cold junction were at $0\,°C$, i.e. $E_{(0-t)}$. Thus a voltage $E_{(0-a)}$ must be added to

Table 2.2
Thermocouple types

Type	Conductors (positive conductor first)	Accuracy	Output for indicated temperature (cold junction at 0 °C)	Service temperature range (°C)
B	Platinum: 30% rhodium alloy Platinum: 6% rhodium alloy	0 to 1100 °C: ±3 °C 1100 to 1550 °C: ±4 °C	1.24 mV at 500 °C	0 to 1500
E	Nickel: chromium/constantan	0 to 400 °C: ±3 °C	6.32 mV at 100 °C	−200 to 850
J	Iron/Constantan	0 to 300 °C: ±3 °C 300 to 850 °C: ±1%	5.27 mV at 100 °C	−200 to 850
K	Nickel: chromium (Chromel)/ Nickel: aluminium (Alumel)	0 to 400 °C: ±3 °C 400 to 1100 °C: ±1%	4.1 mV at 100 °C	−200 to 1100
R	Platinum: 13% rhodium/platinum	0 to 1100 °C: ±1 °C 1100 to 1400 °C: ±2 °C 1400 to 1500 °C: ±3 °C	4.47 mV at 500 °C	0 to 1500
S	Platinum: 10% rhodium/platinum	as type R	4.23 mV at 500 °C	0 to 1500
T	Copper/Constantan	0 to 100 °C: ±1 °C 100 to 400 °C: ±1%	4.28 mV at 100 °C	−250 to 400

Notes: Types B: best life expectancy at high temperatures; E: resistant to oxidizing atmospheres; J: low cost, general purpose; K: general purpose, good in oxidizing atmospheres; R and S: high temperature, corrosion resistant; T: high resistance to corrosion by water.

$E_{(a-t)}$ to correct the output signal:

$$E_{(0-t)} = E_{(a-t)} + E_{(0-a)}. \qquad (2.6)$$

The voltage $E_{(0-a)}$ is called the *cold junction compensation voltage*, and it is provided automatically by the circuit of Fig. 2.6 which includes a thermistor R_t. R_1, R_2 and R_3 are temperature-stable resistors. The bridge is first balanced with all the components at $0\,°C$. As the ambient temperature is changed away from $0\,°C$ an unbalance voltage will appear across AB. This voltage is scaled by selecting R_t such that the unbalance voltage across AB equals $E_{(0-a)}$ in eqn (2.6).

2.3.2 Multiple thermocouple arrangements

Several thermocouples may be connected in series or parallel as shown in Fig. 2.7 to achieve useful functions. The series arrangement (Fig. 2.7(a)) is used mainly as a means of enhancing sensitivity. All the measuring junctions are held at one temperature, and all the reference junctions at another. This arrangement is often called a *thermopile*, and for n thermocouples gives an output n times as great as that which can be obtained from a single couple. A typical commercially available Chromel/Constantan thermopile has 25 junctions and produces about $0.5\,\mathrm{mV/°C}$.

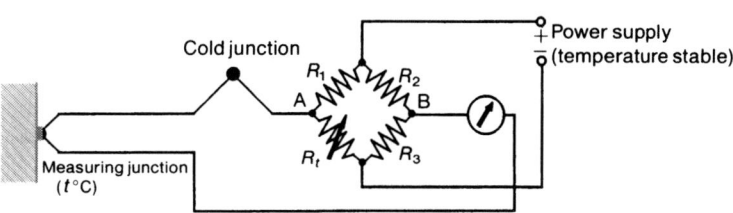

Fig. 2.6 Bridge circuit with cold junction compensation.

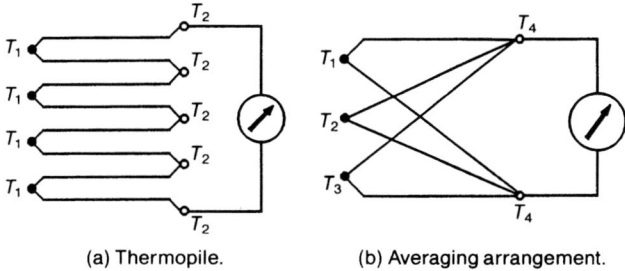

(a) Thermopile. (b) Averaging arrangement.

Fig. 2.7 Multiple junction thermocouples.

The parallel arrangement shown in Fig. 2.7(b) generates the same temperature as a single couple if all the measuring and reference junctions are at common temperatures. If the measuring junctions are at different temperatures and the thermocouples all have the same resistance, the output voltage is the *average* of the individual voltages. The temperature corresponding to the output voltage is the mean temperature (as long as the thermocouples are linear over the measurement range).

2.4 Bimetallic temperature sensors

Bimetallic strips are made by bonding together two metals with different coefficients of thermal expansion. Typical materials used are brass and Invar. As the temperature of the bimetallic component changes the brass side expands or contracts more than the Invar, resulting in a change of curvature.

If two metal strips A and B with coefficients of thermal expansion α_A and α_B are bonded together as shown in Fig. 2.8, a temperature change will cause differential thermal expansion to occur. If it is unrestrained the strip will deflect to form a uniform circular arc. Analysis [3] gives the radius of curvature ρ as:

$$\rho = \frac{t[3(1+m)^2 + (1+mn)\cdot(m^2 + 1/(mn))]}{6(\alpha_A - \alpha_B)\cdot(T_2 - T_1)\cdot(1+m)^2}, \tag{2.7}$$

where ρ is the radius of curvature, m is the thickness ratio, t_B/t_A, t_A, t_B are the thicknesses of strips A and B, t is the total strip thickness, n is the ratio of elastic moduli, E_B/E_A and $T_2 - T_1$ is the temperature rise. In most practical cases $t_B/t_A \approx 1$ and $(n+1)/n \approx 2$, giving

$$\rho \approx \frac{2t}{3(\alpha_A - \alpha_B)\cdot(T_2 - T_1)}. \tag{2.8}$$

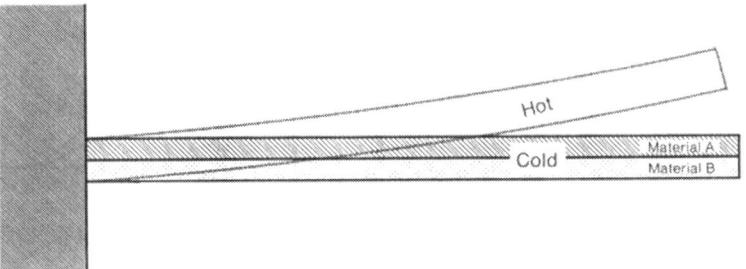

Fig. 2.8 Bimetal strip.

Equations (2.7) or (2.8) can be combined with the appropriate mechanics of solids expressions to calculate the deflections of most shapes of bimetal component, or of the forces developed by partially or completely restrained structures.

Bimetallic devices are used for low-cost, low-accuracy temperature sensing, and as 'on–off' elements in temperature control systems. They are often used as overload cutout switches for electric motors.

2.5 PN junction sensors

PN junctions in silicon have become popular as temperature sensors due to their very low cost. Figure 2.9 shows the forward bias characteristic of a silicon diode. It is well-known that a voltage V_f has to be applied across the junction before a current will flow. For silicon V_f (which is often termed the *diode voltage drop*) is of the order of 600–700 mV. V_f is temperature dependent, and is very nearly linear over the temperature range from −50 to +150 °C. The voltage V_f has a temperature characteristic which is essentially the same for all silicon devices of about −2 mV/°C.

A typical circuit is shown in Fig. 2.10. Bipolar transistors are often used instead of diodes, with the base and collector terminals connected together. For the best performance a constant current should flow through the diodes, but in practice the error incurred by driving the circuit from a constant voltage is small.

There is one major disadvantage to using diodes as temperature sensors in control applications, which is that they are not fail-safe. If a diode temperature sensor is used to control, say, a heater, any breakage of the diode wires will be interpreted by the controller as low temperature. More power will then be applied to the heater, resulting in an uncontrolled runaway.

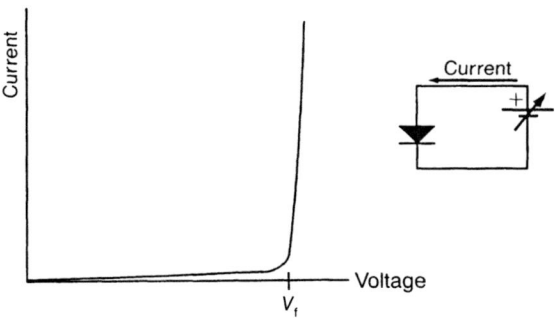

Fig. 2.9 Forward bias characteristic of silicon diode.

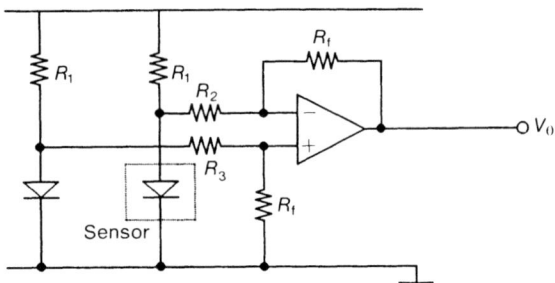

Fig. 2.10 PN junction temperature transducing circuit.

2.6 Liquid crystal temperature sensors

A number of liquids (mainly organic) can be made to exhibit an orderly structure, in which most or all of the molecules are aligned in a common direction. The structure can be altered by electric or magnetic fields. Most people are familiar with the liquid crystal displays used in watches and calculators which use compounds sensitive to electric fields. Less well-known, however, is the fact that some liquid crystal materials are temperature sensitive [3]. One application which may be familiar is the use of cholestoric compounds as medical thermometers, in the form of a flexible plastic strip which is pressed against the skin to measure its surface temperature. The compounds involved have molecular structures similar to cholesterol, and for this reason are called 'cholestoric'. Cholestoric liquids form a helical structure, and this gives them their optical properties. The plane of polarization of light passing through the compound may be rotated by three or more times per millimetre of path length.

The structure can be enhanced by confining the cholestoric liquid between two parallel sheets of a suitable plastic. The choice of polymer for the plastic is based on two requirements. These are first, that it must be optically transparent, and second, that it is sufficiently chemically active to bond to the liquid crystal molecules adjacent to the polymer and maintain their axes in the correct orientation.

When used for temperature measurement the liquid crystal is confined between two polymer sheets as described above, a few tens of micrometres apart. The surface of one plastic sheet is given a reflective coating as shown in Fig. 2.11. In (a) a light ray enters the sandwich and is reflected back. Since the liquid crystal is in its ordered state the reflected ray interferes destructively with the incident light, and the sensor appears opaque.

In Fig. 2.11(b) the liquid crystal has been raised to a temperature at which its ordered structure has broken down due to thermal agitation.

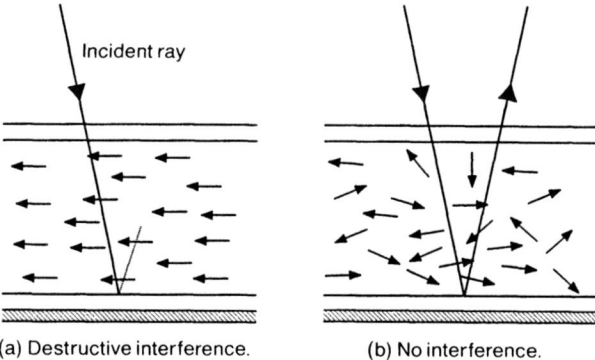

(a) Destructive interference.   (b) No interference.

Fig. 2.11 Liquid crystal temperature sensing.

The temperature at which this occurs depends on the exact molecular structure. It is possible to create compounds tailored to specific temperature changes. If a ray of light enters the sandwich in this condition it is reflected back in the usual way, and the sensor appears to be transparent (the colour of the backing material is usually visible).

If polarized light is used this technique can detect temperature changes as small as 10^{-3} °C. If ordinary white light is used the resolution is of the order of 0.1 °C. The advantages of liquid crystal temperature sensors are that they are relatively cheap, and immune to electromagnetic interference. This last characteristic may make them suitable for applications where RFI problems are too severe for a conventional electronic sensor to be used. It is possible to interrogate a liquid crystal temperature sensor remotely, by means of a fibre optic link.

2.7 Infra-red emission and pyrometry

Most people are familiar with the fact that the amount and the wavelength of radiation emitted by a body are functions of its temperature. This dependence on temperature of the characteristics of radiation is used as the basis of a non-contact temperature measurement technique in which the sensors used are known as *radiation thermometers*.

The total power of radiant flux of all the wavelengths R emitted by a black body of area A is proportional to the fourth power of the temperature of the body in kelvin:

$$R = \sigma A T^4 \qquad (2.9)$$

where σ is the Stefan–Boltzmann constant, which has the value $5.67032 \times 10^{-8}\ \text{W/m}^2\ \text{K}^4$. Most radiation thermometers are based on this law, since if a sensing element of area A at a temperature T_1 receives radiation from an object at temperature T_2, it will receive heat at a rate $\sigma A(T_2)^4$ and will emit heat at a rate $\sigma A(T_1)^4$. The net rate of heat gain is therefore $\sigma A(T_2^4 - T_1^4)$. If the temperature of the sensor is small compared with that of the source T_1^4 may be neglected in comparison with T_2^4.

The above discussion applies to perfectly black bodies with an emissivity ε of unity. 'Real' objects have non-unity emissivities, and a correction must be made for this. The total radiant flux emitted by an object of area A and emissivity ε is:

$$R = \sigma \varepsilon A T^4. \qquad (2.10)$$

The flux R is equal to that emitted by a perfect black body at a temperature T_a, the *apparent temperature* of the body:

$$R = \sigma A T_a^4. \qquad (2.11)$$

Equating 2.10 and 2.11 gives:

$$\sigma \varepsilon A T^4 = \sigma A T_a^4$$

$$\therefore\ T^4 = T_a^4 / \varepsilon$$

$$\therefore\ T = T_a / \sqrt[4]{\varepsilon}.$$

Radiation thermometers generally consist of a cylindrical body made from aluminium alloy or plastic. One end of the body carries a lens, which focuses energy from the target on to a detector within the tube. The lens may be made from glass, germanium, zinc suiphide, sapphire, or quartz, depending on the wavelength. The heat detectors within the instrument are generally thermocouples, thermistors or PN junctions.

2.8 Heat flux gauges

It is sometimes necessary to make local measurements of convective, radiative, or total heat transfer rates. This requirement has led to the development of a class of sensors known as *heat flux gauges*.

One common type of heat flux gauge has the general form shown in Fig. 2.12. Two temperature measuring elements (usually thermocouples) are physically separated by a thermal insulator with known characteristics. When heat energy begins to pass through surface A, thermocouple J1 generates a small voltage. Since the heat has to pass through the thickness of insulating material I to reach the second thermocouple on surface B, a different voltage is generated by J2. The differential voltage

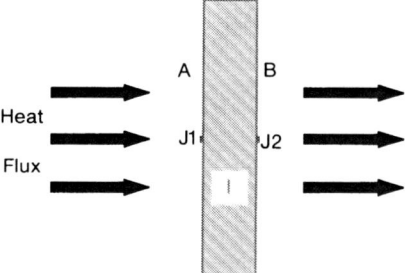

Fig. 2.12 Heat flux gauge.

developed across J1 and J2 is proportional to their temperature difference. If the characteristics of the insulator I are known, the heat transfer rate may be obtained as a function of voltage.

Heat flux gauges of this type are fabricated as a string of thermocouples on a flexible backing. Up to 30 thermocouple junctions (i.e. a thermopile) are placed on each side of the insulator to increase the sensitivity of the device. Heat flux gauges somewhat resemble strain gauges, and are bonded to the surface at which the heat flux is to be measured in a similar fashion. However, strain gauges are usually considered to be disposable, since they are relatively cheap. Heat flux gauges are expensive, and it is usual to try and remove them for re-use. A number of successful removal techniques have been reported [4].

References

[1] *Engineering Thermodynamics, Work and Heat Transfer.* Rogers and Mayhew. 3rd edition. Longman, 1983.
[2] *Investigation of the Properties of Thick-film Thermistor Pastes.* P.N. Dargie. USITT Report No. U104-2. Available from USITT, University of Southampton, Southampton SO5 9NH, UK.
[3] *Measurement Systems Application and Design.* E.O. Doebelin. 4th Edition. McGraw-Hill, 1990.
[4] The use of thin-foil heat flux gauges to determine plug closure in cryogenic pipe freezing. M.J. Burton and R.J. Bowen. In *Proceedings of the 12th International Cryogenic Engineering Conference* (ICEC12), Southampton, 1988.

Displacement sensing

3.1 Introduction

Measurements of the displacement of an object are of fundamental importance in experimental science, and are the basis of measurements of velocity, acceleration, strain, and (by the use of elastic elements) force and pressure. Either translational or rotational displacement measurements may be needed. The principles underlying the operation of many displacement sensors are common to both linear and rotary types. For this reason these two types of measurement have not been treated separately.

3.2 Potentiometers

A potentiometer consists essentially of a resistive element which is provided with a movable contact. In the earliest forms of potentiometer the resistive element or 'track' was made from high-resistance wire, such as Nichrome, wound on to an insulating former of suitable shape. The contact consists of a springy, conducting arm, which is arranged so that it can be moved along the potentiometer track. A variable resistance is thus created between one end of the track and the movable contact. The contact motion can be linear, rotary, or a combination of the two, such as helical movement.

Translational (also called linear[1]) potentiometers are available with strokes from about 5 to 1000 mm. Rotary versions are available with strokes from about $10°$ to as much as 60 turns ($>20\,000°$).

3.2.1 Potentiometer linearity

If the resistance of a potentiometer is linear with respect to its travel, the output voltage e_0 (see Fig. 3.1) is a linear function of displacement x_i when an excitation voltage e_{ex} is provided, the output is open-circuit and

---

[1] The term 'linear' is used somewhat confusingly with respect to potentiometers, and can have two meanings. It is often used to denote a device in which the contact motion is translational. However, it is also frequently used to describe a device in which the resistance is proportional to displacement (rather than being, say, logarithmic). Such a 'linear' potentiometer can of course have a contact which undergoes circular, helical, or translational motion, and hence the confusion arises.

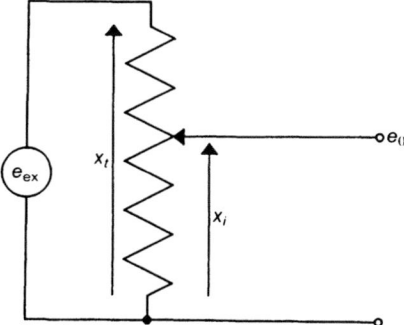

Fig. 3.1 Potentiometric displacement sensor.

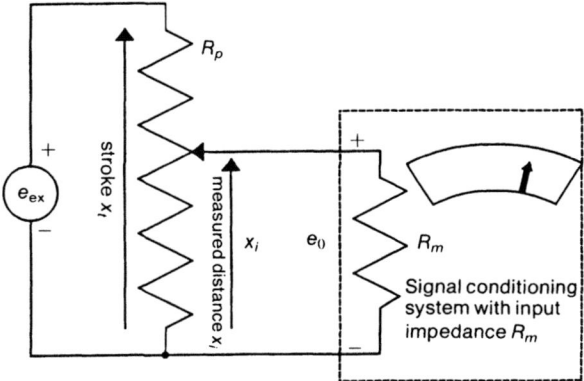

Fig. 3.2 Potentiometer connected to input impedance R_m.

no current is drawn. However, all circuit inputs draw some current, so any signal conditioning arrangement connected to the potentiometer will degrade its linearity to some extent. Figure 3.2 shows a common arrangement, and from simple circuit analysis it can be shown that:

$$\frac{e_0}{e_{ex}} = \frac{1}{(x_t/x_i) + (R_p/R_m)(1 - x_i/x_t)}. \tag{3.1}$$

Under ideal conditions $R_p/R_m = 0$ for an open circuit, and eqn (3.1) becomes:

$$\frac{e_0}{e_{ex}} = \frac{x_i}{x_t}. \tag{3.2}$$

Thus when no current is drawn the input–output relationship is a straight line. In practice $R_m \neq \infty$, and as shown in Fig. 3.3 there is a nonlinear

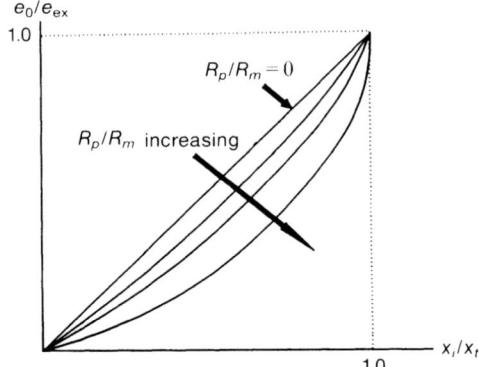

Fig. 3.3 Potentiometer loading effects.

relationship between e_0 and x_i. If $R_p = R_m$, the maximum deviation from linearity is about 12%. If $R_p = 10\%$ of R_m, the error drops to about 1.5%. For values of $R_p/R_m < 0.1$ the position of maximum error is in the region where $x_i/x_t \approx 0.67$, and maximum error is approximately $15 R_p/R_m\%$ of full scale.

To achieve good linearity therefore the input impedance R_m of any circuit connected to a potentiometer should be high compared with the potentiometer impedance R_p, which should be kept as low as possible. Unfortunately this requirement conflicts with the almost invariable need for high sensitivity. Since the output e_0 is directly proportional to the excitation voltage e_{ex}, at first sight it appears to be possible to get any desired output simply by increasing e_{ex}. However, potentiometers have a fixed power rating which is determined by their heat-dissipating capability. If the limiting heat dissipation is H watts, the maximum allowable excitation voltage is:

$$e_{ex}(\text{max}) = \sqrt{HR_p}. \tag{3.3}$$

Thus, a low value of R_p allows only a small e_{ex}, and therefore a reduced sensitivity. The choice of R_p must be a compromise between considerations of loading and sensitivity.

3.2.2 Potentiometer resolution

The resolution of a potentiometer depends on its construction. In a wirewound type the variation in resistance proceeds in small steps as the wiper moves from one turn of resistance wire to the next, as shown in Fig. 3.4. The finest wire spacing that can be achieved in around 25 turns/mm, and thus for wirewound translational devices the best resolution that can be achieved is about $\pm 40\,\mu\text{m}$.

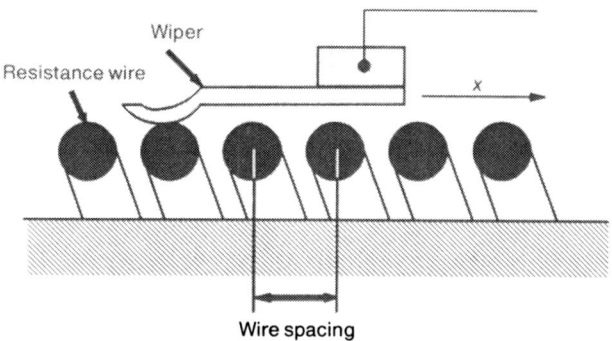

Fig. 3.4 Wirewound potentiometer resolution depends on wire spacing.

The resolution of *carbon-film*, *cermet* (a mixture of ceramic and metallic materials) or *conductive plastic* (a mixture of plastic resin and a conducting powder) resistance elements is higher than that of wirewound devices, but they are not so hard-wearing. This type of potentiometer is often described as having infinite resolution, since the resistance element presents a smooth surface to the wiper. The resolution cannot actually be measured and quantified, since the deviations of e_0 from the ideal straight line are random (in contrast to the repeatable output 'steps' of a wirewound device). The quantity usually cited in an attempt to specify the output smoothness is the ratio of the peak amplitude of the random variations to e_{ex}, with 0.1% being a typical value.

A further problem with non-wirewound resistance elements is that they cannot tolerate high excitation currents. As we have seen in the previous section, this leads to a reduction in sensitivity.

A compromise solution is offered by the so-called *hybrid potentiometer*, in which a layer of conductive plastic is applied on top of a wirewound track. This approach combines the best features of both types of potentiometer, but at an increased cost.

3.2.3 Electrical noise problems in potentiometers

Electrical noise in a potentiometer arises from output voltage fluctuations due to contact slider bounce, dirt, and contact or track wear. The term 'noise' is often used to include the effects of resolution. It should not be forgotten that in dynamic applications a potentiometer can cause significant mechanical loading due to the inertia and friction of its moving parts. This can affect the characteristics of the motion being measured.

In a wirewound potentiometer the sliding contact may 'bounce' at certain speeds as it passes over the turns of wire, causing intermittent contacting. This effect can be particularly severe if the speed and wire spacing are such that an exciting force occurs at or close to the fundamental frequency of the spring-loaded contact arm. A solution to this problem which is sometimes adopted is to use a contact divided into two or more parts, each with a different resonance frequency. Thus, if one part is resonating the other is still likely to make a good contact.

3.3 Inductive displacement transducers

Inductive position transducers do not suffer from the problems associated with a sliding contact, since they are inherently non-contact devices. The resolution available from a good-quality linear variable differential transformer (or LVDT—see later) is equal to that obtained from a potentiometer. However, for many applications inductive sensors suffer from one inherent disadvantage: they are essentially AC devices, and cannot be run from DC battery supplies without extra complications.

3.3.1 Variable reluctance transducers

To understand how inductive displacement transducers work, the concept of a *magnetic circuit* is required. In an electric circuit an electromotive force or voltage V drives current I through a resistance R. From Ohm's law the voltage and current are related by:

$$V = I \times R. \tag{3.4}$$

In a magnetic circuit, such as the example shown in Fig. 3.5(a), a current i passes through a coil of n turns, which is wound on to a loop of ferromagnetic material. By analogy we can regard the coil as a source of *magnetomotive force*, (mmf), which drives a flux ϕ through the magnetic circuit.

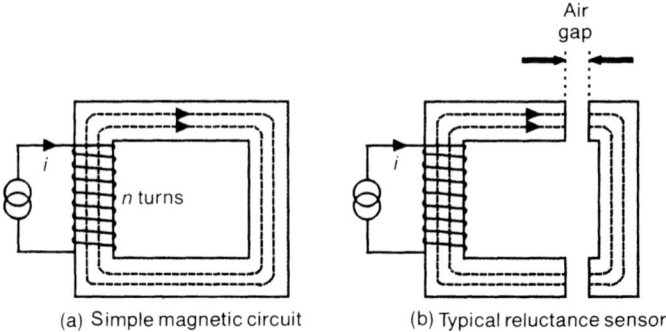

(a) Simple magnetic circuit   (b) Typical reluctance sensor

Fig. 3.5 Variable reluctance sensing.

The corresponding equation is:

$$\text{mmf} = \text{flux} \times \text{reluctance} = \phi \times \mathfrak{R}, \qquad (3.5)$$

where $\mathfrak{R}$ is the *reluctance* which limits the flux in the magnetic circuit, just as resistance limits the current in an electric circuit. In the example shown in Fig. 3.5 the mmf is $n \times i$, so in this case the magnetic circuit flux ϕ linked by a *single turn* of the coil is given by eqn (3.6):

$$\phi = \frac{ni}{\mathfrak{R}} \text{ (webers)}. \qquad (3.6)$$

The total flux N linked by the *entire* coil of n turns is:

$$N = n\phi = \frac{n^2 i}{\mathfrak{R}}. \qquad (3.7)$$

By definition the self-inductance L of the coil is the total flux per unit current [1]. Therefore:

$$L = \frac{N}{i} = \frac{n^2}{\mathfrak{R}}. \qquad (3.8)$$

Equation (3.8) allows us to calculate the inductance of a sensing element given the reluctance of the magnetic circuit. The reluctance $\mathfrak{R}$ of a magnetic circuit is given by:

$$\mathfrak{R} = \frac{\mathscr{L}}{\mu \mu_0 A}, \qquad (3.9)$$

where $\mathscr{L}$ is the total length of the flux path, μ is the relative permeability of the magnetic circuit material, μ_0 is the permeability of free space ($= 4\pi \times 10^{-7}$ H/m), and A is the cross-sectional area of the flux path.

Figure 3.5(b) shows an arrangement similar to that of Fig. 3.5(a), except that the magnetic core has been separated into two parts by an air gap of variable width. The total reluctance of the circuit is now the sum of the reluctances of the two parts of the core, and of the air gap. The relative permeability of air (μ_{air}) is close to unity, while that of the core material can be several thousand times greater. Thus, the presence of the air gap causes a large increase in the circuit reluctance, and a corresponding decrease in flux and inductance. It is this effect which is used in sensor construction, since a small change in the width of the air gap causes an easily measurable variation in inductance. In engineering variable reluctance sensors are most frequently used for sensing rotation rate, where (as shown in Fig. 3.6) the sensor is placed close to a toothed wheel. Movement of the ferromagnetic teeth past the device causes a variation in the coupling between the coils. After suitable signal conditioning the

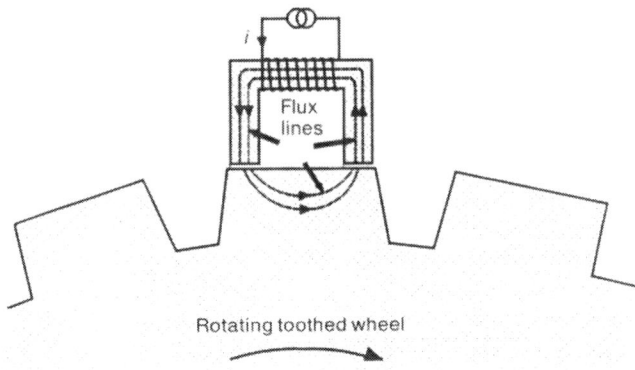

Fig. 3.6 Variable reluctance rotation rate sensor.

output signal is produced with a frequency which is linearly related to rotation rate.

3.3.2 Variable coupling transformers: LDTs and LVDTs

A simple form of inductive displacement transducer known as a *linear displacement transducer* or LDT may be made by winding a pair of coils (or a single centre-tapped coil) on to a hollow cylindrical former, and allowing a ferromagnetic plunger to move along the axis as shown in Fig. 3.7(a). The plunger and coils have the same length, d. As the plunger is moved the inductances vary. The two inductances $L1$ and $L2$ are normally connected in a bridge circuit with a pair of balancing resistors R, followed by an amplifier as shown in Fig. 3.7(b). If the inductances are both L when the plunger is in the central position, and if the plunger and coil lengths are all d, a displacement of δd will produce opposing inductance changes $\pm\delta L$. Ideally therefore $\delta L/L = \delta d/(d/2)$, and the corresponding bridge output will be $(e_{ex}/2)(\delta L/L)$.

A problem with the device described above is that the output is only linear over a limited region, when the plunger is close to the centre. An improved performance is obtained from the slightly more complex device known as a *linear variable differential transformer* or LVDT. An LVDT is a transformer with a single primary winding and two secondaries, wound end-to-end on a tubular former as shown in Fig. 3.8. A ferromagnetic plunger (often a ferrite rod) moves inside the tubular former, and varies the coupling between the primary and the secondaries. Both linear and rotational versions are available as shown in Fig. 3.8.

The centre (primary) coil is fed from an AC excitation supply, and induces voltage across the two outer (secondary) coils. The induced

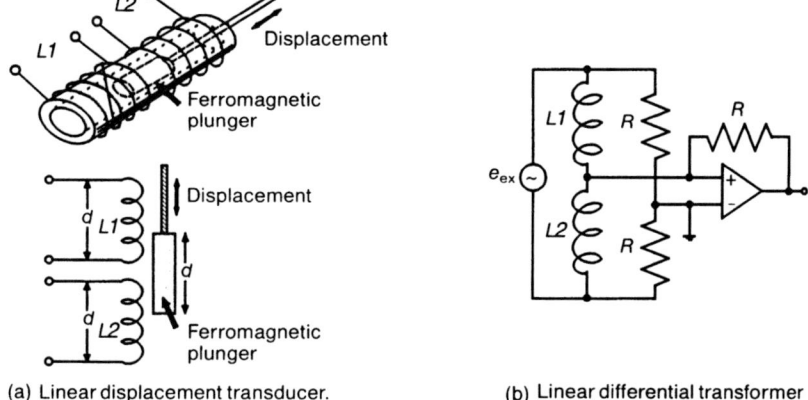

(a) Linear displacement transducer.

(b) Linear differential transformer connected to bridge circuit.

Fig. 3.7 Linear differential transformer used as displacement sensor.

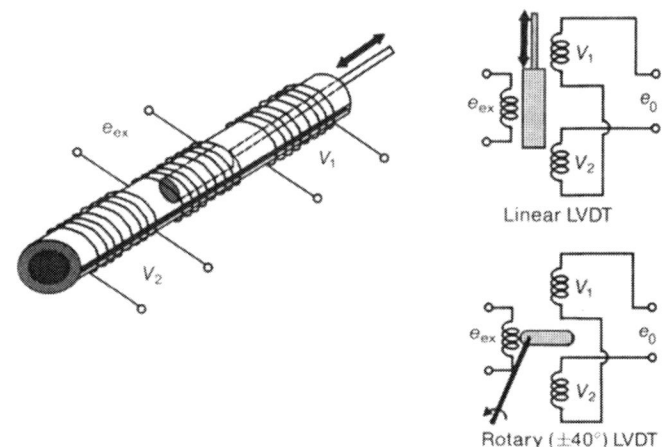

Linear LVDT

Rotary ($\pm 40°$) LVDT

Fig. 3.8 Linear variable differential transformer.

voltages have equal magnitudes when the plunger is positioned symmetrically. If the excitation voltage $e_{ex} = V_s \sin(2\pi f_s t)$, the output e_0 is given by the difference $V_1 - V_2$ of the voltages induced in the two secondaries. The secondaries are normally connected in *series opposition*, as shown in Fig. 3.8, so that the output voltage is:

$$e_0 = V_1 - V_2 = V_{out} \sin(2\pi f_s t + \phi). \tag{3.10}$$

An LVDT is usually excited by a sinusoid up to around 24 V in amplitude, with a frequency in the range 50 Hz to 25 kHz. With the series opposition arrangement a null position exists when the plunger is centrally placed for which the output $e_0 \approx 0$. A displacement of the plunger away from the null position increases the coupling (mutual inductance) between the primary and one secondary, while decreasing the coupling for the other secondary. The amplitude of e_0 is almost linear with respect to displacement for a considerable range either side of the central 'null' position as shown in Fig. 3.9.

The output, e_0 also undergoes a 180° phase shift when the plunger is moved past the central position. An LVDT is usually used in conjunction with a signal conditioning circuit which converts the output signal into DC. The conversion is carried out in such a way that the regions A and B in Fig. 3.9 are distinguished, where the amplitude is the same but the phase difference is 180°. A *phase-sensitive demodulator* is used [2] to sense the phase difference, and gives a negative output voltage for displacements in the A region and a positive output for B. These signal conditioning circuits are sometimes incorporated within the casing of the LVDT, and operation of the device is then 'transparent' to the user: a DC supply voltage is used to energize the device, and a DC position signal is returned.

Nonlinear effects occur at the extremes of travel as shown by regions D and E on Fig. 3.9. A typical nonlinearity figure for an LVDT is ±1% over the whole range of the device.

LVDTs are available to cover ranges from ±0.25 mm to ±500 mm. The useful frequency range is limited by the inertia of the moving parts of the device. For many applications the main problem with an LVDT is the cost. However, LVDTs are very frequently used in laboratory and prototype development work.

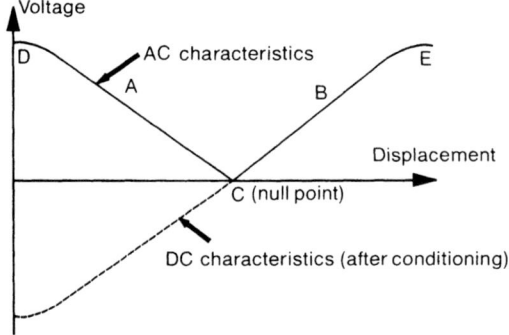

Fig. 3.9 Amplitude of LVDT output voltage as a function of displacement.

3.3.3 Eddy current displacement transducers

In appearance an eddy current transducer resembles a variable-reluctance sensor, but with one important difference: the 'target' used by a variable-reluctance device has to be ferromagnetic, while the object being sensed by an eddy-current probe simply has to be electrically conducting. Thus, eddy-current probes can be used in more situations than variable-reluctance devices.

An eddy-current probe consists of a pair of coils, one active, which is influenced by the proximity of a conducting target, and a second balance coil which is included to complete a bridge circuit and provide temperature compensation.

The bridge is excited by a high frequency (typically 1 MHz) AC signal, and magnetic flux lines from the active coil pass into the surface of the conducting target. Eddy currents are produced, largely at the surface, and diminish away from it. They are negligibly small a short distance below the surface.

As the target moves with respect to the probe the eddy currents vary, which changes the impedance of the active coil. A bridge imbalance results, which is related to the distance between the active coil and the target. This unbalance signal is filtered and rectified to provide a DC voltage signal. The signal is inherently nonlinear, and linearization circuits are sometimes employed to remedy this defect.

Probes are commercially available to cover the ranges from ± 0.25 mm to ± 30 mm, with a maximum resolution of 10^{-4} mm. The recommended measuring range for a given probe begins at a 'standoff' distance which is normally about 25% of the range. Thus, a probe with a ± 1 mm range is mounted with a static target/probe distance of between 0.5 and 1.5 mm.

Targets are not supplied with eddy-current probes, since an existing machine part is normally used. The motion of a non-conducting target can be sensed if a piece of conducting foil is fixed to the surface. Adhesive-backed aluminium foil tape is commercially available for this purpose.

Flat targets should be the same diameter as the probe, or larger if possible. Curved targets (such as circular shafts) behave like flat surfaces if the shaft diameter exceeds five transducer diameters. Special four-probe systems are available for measuring the orbital motion of rotating shafts.

3.4 Capacitive displacement transducers

Capacitance occurs when two electrically conducting components are separated by a dielectric (or non-conducting) medium. When a voltage V is applied across the capacitor, equal and opposite charges $\pm Q$ appear on the two conducting components. The ratio of charge to voltage defines the capacitance of the device:

$$C = \frac{Q}{V}. \tag{3.11}$$

If the capacitor consists of two parallel conducting plates of area A, separated by a distance d, then the capacitance C is

$$C = \frac{\varepsilon_0 \varepsilon_r A}{d} \text{ (in farads (F))} \tag{3.12}$$

where ε_0 is the permittivity of free space (a vacuum) which has a value of 8.85×10^{-12} F/m, and ε_r is the relative permittivity of the dielectric material separating the conducting plates. For dry air $\varepsilon_r = 1$.

Capacitance is therefore a function of geometrical parameters (A and d) as well as the material property ε_r. Any change in A, d, or ε_r causes a corresponding change in the capacitance as shown by eqn (3.12). This fact is used in the design of capacitive displacement sensors, several versions of which are shown in Fig. 3.10. Capacitive sensing is common where accurate measurement of small displacements is required, such as in pressure and acceleration sensing.

Single or differential outputs can be used, although in practice it is often inconvenient to employ only two plates since the output is non-linear as discussed below. A three-plate differential arrangement is normally used. There are several common geometries for both translational

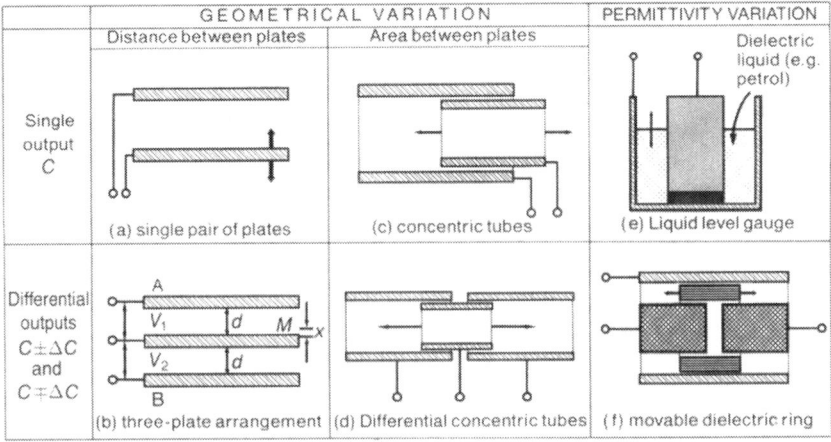

Fig. 3.10 Capacitive displacement transducers.

and rotary displacement sensing, some of which are shown ing Fig. 3.10. The nominal capacitance of this type of sensor is generally below 1000 pF (1 nF). To reduce the impedance levels and to allow the determination of high-speed movement, the frequency of the signal used for measurement is usually 100 kHz or greater. The transducer impedance at this frequency is still quite high, so the insulation resistance must also be high to avoid 'shunting' the transducer unduly and reducing its sensitivity. A charge amplifier is almost invariably used. Capacitive displacement transducers are normally used to measure displacements of less than 1 mm.

Capacitive sensors are often unreliable when used in conditions where the humidity may vary. This is probably why their use in engineering is largely confined to pressure and acceleration sensors, where the capacitive element can be sealed in a vacuum or enclosed in an atmosphere of dry gas.

A sensor in which the plate area A or separation d is varied requires a physical connection between the motion being sensed and the moving part of the capacitor (see Fig. 3.10). Transducers based on permittivity variation ε_r on the other hand do not require any physical connection, and this has led to their use being preferred in some circumstances.

3.4.1 Linearity of capacitive displacement transducers

In a simple two-plate arrangement such as that shown in Fig. 3.10(a), the variation of C as d alters is hyperbolic and can only be treated as linear for small excursions. In this case the sensitivity $\Delta C/\Delta d$ is proportional to $1/d^2$.

If a three-plate differential arrangement such as that shown in Fig 3.10(b), is used, a linear output can be obtained. Referring to the diagram, it is apparent that the fixed plates A and B and the movable plate M form a pair of capacitors with values $C1$ and $C2$. When M is centrally positioned midway between A and B, a distance d from each, $C1 = C2$. If M is moved a distance x towards A the capacitances become:

$$C1 = \frac{\varepsilon_0 \varepsilon_r A}{(d+x)} \quad \text{and} \quad C2 = \frac{\varepsilon_0 \varepsilon_r A}{(d-x)}. \tag{3.13}$$

If a voltage V_{ex} is applied across AB, the output V_{out} is given by

$$V_{out} = V1 - V2 = V_{ex} \left[\frac{C2}{(C1+C2)} - \frac{C1}{(C1+C2)} \right] = V_{ex} \cdot \frac{x}{d}. \tag{3.14}$$

The three-plate arrangement is therefore linear, with V_{out} proportional to x, while the sensitivity V_{out}/x is inversely proportional to d.

3.5 Optical motion sensors

Angular and translational (linear) motion are often measured optically using pulse counting methods. Optical position transducers are becoming popular for the following reasons:

- They are inherently digital.
- They are immune from electrical interference.
- No mechanical connection to the sensing element is made.
- They can be low-cost, especially if plastic optical components are used.

Their main drawbacks are that they are relatively fragile, and that their performance suffers badly if dirt is allowed to get onto the optical components. For these reasons they are largely restricted to applications where the sensor can be mounted in a relatively clean environment, and does not experience extremes of temperature.

3.5.1 Angular optical encoders

The simplest form of angular encoder consists of a slotted or striped disk, a light source and a photo-sensitive detector. The slots in the disk 'chop' the light falling on the detector. If a reflective disk with matt segments is used, the light source and detector can be placed on the same side of the disk. If a slotted version is used they are placed on either side. The resulting pulse train indicates rotational speed. If the disk rotates at n revolutions per minute (rpm), the output pulse rate is $nS/60$ Hz, where S is the number of slots in the disk. Simple devices such as this cannot give any indication of the direction of rotation, nor in general can they measure angular position unless a starting reference-point is provided and pulse counting techniques used.

The difficulty over sensing the direction of rotation may be overcome if two tracks and two sensing heads are used. The tracks are positioned so that their electrical outputs have a phase displacement which lags or leads, depending on the direction of rotation.

The transducers described above are usually referred to as *incremental angular encoders*, since unless the start-up point is known it is very difficult to estimate the absolute position. This difficulty is overcome by *absolute angular encoders* such as the 4-bit examples shown in Fig. 3.11. These devices consist of a number of concentric tracks containing a pattern of opaque and transparent (or reflective) sections, such that a unique binary code can be read from the patterns for any angle. The patterns are read by means of photocells as before.

If natural binary coding is used, as on the disk shown in Fig. 3.11(a), a potentially large error can occur if a readout is attempted when the photocells lie along a line midway between two sectors of the disk. It can be seen from the diagram that, for some positions, a number of tracks change state simultaneously. (These are known as the *cardinal transitions*.) This problem does not arise if Gray code is used, as shown on Fig. 3.11(b), since only one bit changes state at each sector. It is therefore impossible to generate an output which is in error by more than one bit.

Absolute encoders are available with 8, 10, 12, or 16 tracks, but they become increasingly expensive and inconveniently large as the number of tracks increases. The resolution obtained from an n-track absolute encoder is $360/(2^n)$ degrees. Thus, an eight-track device, for example, is divided into 256 ($=2^8$) sectors, and can resolve to within $360/256 = 1.41°$.

An alternative approach which dispenses with the need for n tracks to obtain an accuracy of $360/(2^n)$ degrees is to use a single track containing a maximal-length binary sequence of 2^n digits for n bits. Such a sequence contains all the possible combinations of n bits exactly once, and every angular readout (determined by n adjacent bits) is unique. A separate clock track is often used with this scheme. Figure 3.12 shows a simple 4-bit version of this arrangement.

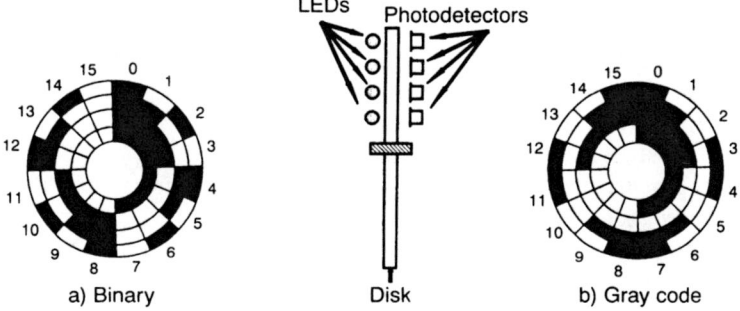

Fig. 3.11 Binary and Gray code absolute angular encoders.

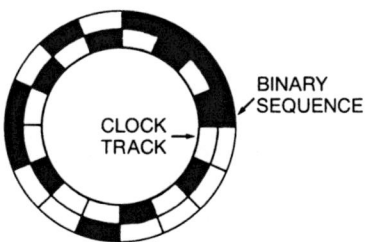

Fig. 3.12 Absolute encoder with 4-bit maximal-length binary sequence.

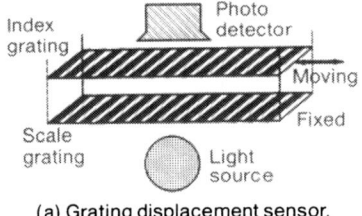

(a) Grating displacement sensor.   (b) Two-phase index grating.

Fig. 3.13 Translational optical sensors.

3.5.2 Translational optical encoders

The arrangements described above for angular displacement sensing can also be applied to translational motion. A device commonly used in high-precision displacement sensing consists of two diffraction gratings as shown in Fig. 3.13(a). One part (the scale grating) is fixed, while the other (the index grating) is attached to the moving object and slides over the fixed grating. As the index grating moves the appearance of the device alternates between light and dark, and this change is readily sensed by means of a photocell. The photocell does not have to be the same size as the lines on the grating, and indeed there are advantages in making it much larger, since any errors introduced by imperfections in individual lines are automatically averaged out.

Just as for angular position sensing, a single photocell feeding a counting or frequency measuring system gives no indication of the direction of movement. If this is required the index grating is made in two parts as shown in Fig. 3.13(b), where the two sets of lines are displaced by 1/4 of a line spacing. The resulting quadrature signals are often used to feed a direction detector and an up/down counter.

Absolute translational encoders are also available. These operate in much the same way as the angular versions, and consist of a number of parallel tracks containing a pattern of opaque and transparent (or reflective) sections, such that a unique binary code can be read from the patterns for any angle. Once again Gray code is used to reduce errors.

3.6 Ultrasonic displacement transducers

The term *ultrasound* refers to sound waves generated at frequencies higher than the human ear can detect, i.e. at frequencies above about 18 kHz. Ultrasonic waves obey the same basic laws of wave motion as lower-frequency acoustic waves [3]. They have, however, the following advantages:

- Higher frequencies imply shorter wavelengths, which means that diffraction does not occur so readily. It is easier to produce a directed and focused beam of ultrasound than it is for 'ordinary' sound.
- Ultrasonic waves pass easily through the metal walls of a structure. This means that for applications such as fuel tank level sensing the measurement system may be mounted externally.

Ultrasound has been used for distance measurement or rangefinding for many years. Examples with which the reader will be familiar include marine sonars, medical ultrasound imaging equipment, and defect location systems for use in structural materials such as steel and concrete. In each case brief pulses of ultrasound are emitted, and the range of the object under investigation is deduced by timing the echo.

The possibility of using an ultrasonic echo method for obstacle detection was first suggested by L.F. Richardson at the time of the *Titanic* disaster in 1912 [4]. The idea was developed intensively during the First World War with the aim of detecting enemy submarines, and the first practical system was developed by P. Langevin. Figure 3.14 shows a diagram of Langevin's quartz ultrasonic transducer, which was used both as a transmitter and receiver. A beam of ultrasound was propagated vertically

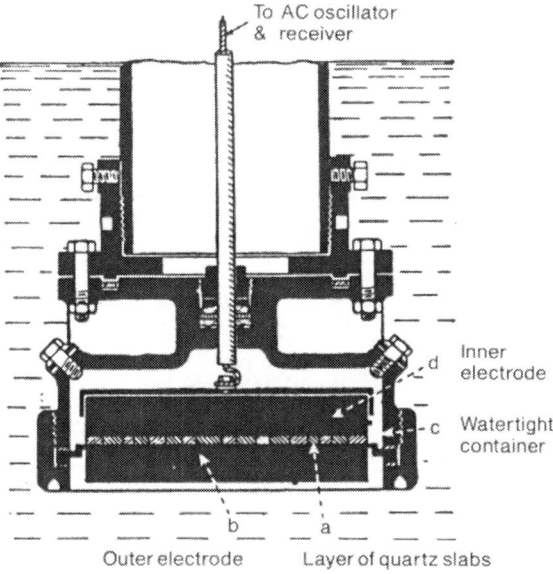

Fig. 3.14 Langevin's ultrasound transducer.

downwards into the sea and reflects from the bottom, or from any intervening object such as a submarine. Langevin's work forms the basis of all modern sonar (<u>so</u>und <u>na</u>vigation and <u>r</u>anging) systems, both for military uses and for applications such as navigation and fish-detection.

Sonar was the first engineering application of ultrasound, and its development preceded that of radar by more than 30 years. The development of radar led to a number of improvements in electronic technology, many of which were in turn applied to ultrasound systems. One of the most important of these developments was the possibility of using a *phased array* [4] to control the direction of the ultrasound beam.

Radar cannot be used under water because (since seawater is conducting) electromagnetic waves are very rapidly attenuated. The attenuation of compression waves in water, however, is very low, and submarine sonars can operate over distances of several kilometres. The attenuation of ultrasound in air is much greater than in water, but practical ultrasonic ranging systems can still be built which operate satisfactorily over ranges of tens of metres. The attenuation of ultrasound in air is greatest at high frequencies, so in airborne applications frequencies of the order of tens of kHz are used. For example, airborne ultrasound is used to provide automatic focusing in some cameras, and for this application tone bursts with a frequency between 20 and 25 kHz are used.

In automotive engineering ultrasound has been used to measure the *ride height* of a vehicle for adaptive suspension control, to monitor the road profile of a vehicle in advance of the wheels, again for adaptive suspension control, and for local collision warning and avoidance systems (often called parking or reversing aids).

A number of aircraft fuel gauging systems use ultrasound for tank level sensing. This development was introduced because it offers the possibility of non-invasively (and therefore safely) measuring the amount of liquid in a fuel tank. Sophisticated systems have been developed in which measurements from a number of transducers are integrated to give a reading which is independent of the motion of the fuel ('sloshing').

3.7 Hall effect motion sensors

If a conducting or semiconducting material carrying a current is placed in a magnetic field B, which is normal to the direction of current flow I as shown in Fig. 3.15, a voltage V is produced across the width of the conductor. The effect is greatest in semiconductors, and these are normally used in commercial Hall effect devices. Briefly, the Hall voltage appears because the magnetic field causes the current carriers to follow a

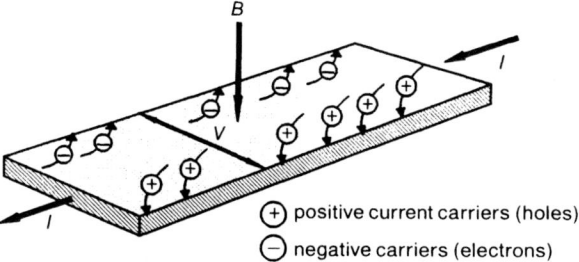

Fig. 3.15 The Hall effect in a (semi)conductor.

curved path as shown in the diagram. An excess of current carriers appears along one edge of the device, while a shortage occurs at the opposite side. This charge imbalance gives rise to the Hall voltage, which remains as long as the magnetic field is present and as long as the current is maintained. For a rectangular (semi)conductor of thickness t the Hall voltage V is given by:

$$V = \frac{K_H B I}{t},\qquad(3.15)$$

where K_H is the Hall coefficient for the material, which depends on the charge mobility and resistivity of the conductor. Iridium antimonide is widely used in Hall effect devices, for which $K_H = 20\ V/T$.

Hall effect transducers may be used for non-invasive current measurement, for magnetic field measurement, or for distance, velocity, or proximity sensing. This last group of applications is the most common in mechanical engineering, where Hall probes are often used to sense rotation rate by means of a toothed wheel.

3.7.1 Hall probe rotation rate sensors

Hall pickups for sensing rotation rate sometimes use a permanent magnet fixed to the moving object. More frequently a ferrous target (such as a toothed wheel) is used, whose approach changes the reluctance of an internal magnetic circuit which the Hall probe 'feels'. This kind of system is frequently used to sense wheel speed for automotive anti-lock braking (ABS) and traction control systems [5]. A typical arrangement is shown in Fig. 3.16.

3.7.2 Hall probe displacement measurement systems

Hall effect transducers give an output voltage which is linearly and repeatably related to the magnetic field intensity. If a suitably-shaped permanent

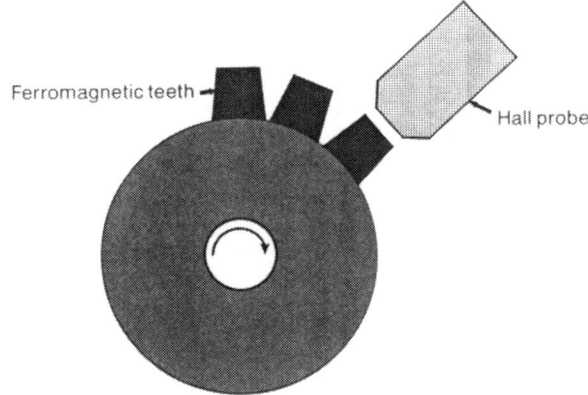

Fig. 3.16 Hall probe used to measure rate of rotation of toothed wheel.

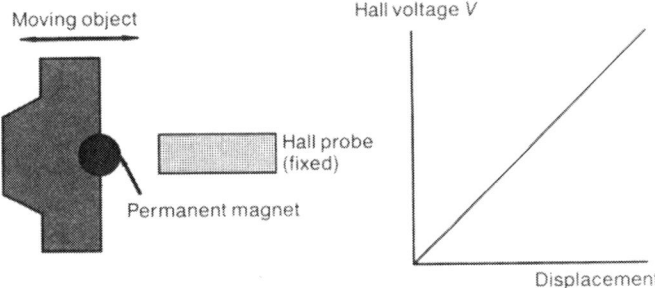

Fig. 3.17 Hall transducer used for displacement measuring.

magnet is used the Hall probe output can be made inversely proportional to the distance from the permanent magnet. This effect can be exploited to give a non-contact displacement sensor which is reasonably accurate (typically ±1%). A permanent magnet is fixed to the moving object as shown in Fig. 3.17, while the Hall probe is fixed. Distance measurements can be carried out using this approach within the range 0–20 mm. However, care should be taken to provide adequate magnetic shielding, since errors will be introduced if the sensor is exposed to transient magnetic fields. This is probably the reason why relatively few uses of this technique have been reported. One example is discussed in reference [6], in which a Hall effect accelerometer is described. A permanent magnet acts as the seismic mass in this device, and the Hall probe senses its motion.

References

[1] *Electromagnetism*, by I.S. Grant and W.R. Phillips. Wiley.

[2] *Instrumentation reference book*, edited by B.E. Noltingk. Butterworths, 1990.

[3] *Acoustics for engineers*, by J.D. Turner and A.J. Pretlove. Macmillan, 1991.

[4] *Ultrasonics*, by A.P. Cracknell. Wykeham Publications, 1980.

[5] A differential Hall IC for geartooth sensing, by R. Podeswa and U. Lachmann. Paper C391/059, in *Proceedings of the 7th International Conference on Automotive Electronics*, I.Mech.E, 1989.

[6] A Hall effect accelerometer, by R.E. Bicking. Paper C391/036, in *Proceedings of the 7th International Conference on Automotive Electronics*, I.Mech.E, 1989.

Velocity and acceleration transducers

4.1 Introduction

Displacement, velocity, and acceleration are intimately linked together, and measurements of one quantity may be used to estimate the others. This is fortunate, since the device almost always used for vibration measurement is the accelerometer, which measures acceleration. However, estimates of displacement or velocity are required almost as often as those of acceleration.

The relationship between the peak values of acceleration, velocity, and displacement is easily demonstrated. Suppose, for example, an object undergoes harmonic oscillation, and experiences a displacement u such that:

$$u = A \cdot \cos(\omega t). \qquad (4.1)$$

The corresponding velocity v (obtained by differentiating) will be:

$$v = \frac{du}{dt} = -A\omega \sin(\omega t). \qquad (4.2)$$

Differentiating again to obtain the acceleration a gives:

$$a = \frac{d^2 u}{dt^2} = -A\omega^2 \cos(\omega t). \qquad (4.3)$$

Considering only the absolute values, we can see that:

$|u| = A,$

$|v| = A\omega,$ and

$|a| = A\omega^2.$

So if, for example, the vibration of an object is measured and found to have a peak acceleration of 15.8 m/s² at 200 Hz, the peak velocity will be:

$$|v| = \frac{|a|}{\omega}$$

$$= \frac{15.8}{2\pi \times 200} = 0.0126 \, \text{m/s}.$$

Similarly, the peak displacement will be:

$$|u| = \frac{|a|}{\omega^2} = \frac{15.8}{(2\pi \times 200)^2} = 0.1 \, \text{mm}.$$

Some forms of velocity transducer use the fact that relative motion between an electrical conductor and a magnetic field, such that the conductor cuts through the field lines, generates a voltage across the ends of the conductor. Magnetic velocity sensors can be divided into two types, according to whether a moving coil or a moving magnet is used.

Other transducers (accelerometers and seismometers) use the effect of Newton's second law, which states that when a mass is accelerated a force is produced according to the equation:

$$\text{force} = \text{mass} \times \text{acceleration}.$$

4.2 Accelerometer and seismometer theory

Most vibration sensors are based on the damped mass-spring system shown in Fig. 4.1. Depending on whether the frequency to be measured, ω, is higher or lower than the resonance frequency of the sensor, ω_n, displacement, velocity or acceleration may be measured. Transducers of this type operate by sensing the motion of the suspended mass m relative to the case.

The equation of motion for the system of Fig. 4.1 is:

$$m\ddot{x} = -c(\dot{x} - \dot{y}) - k(x - y), \tag{4.4}$$

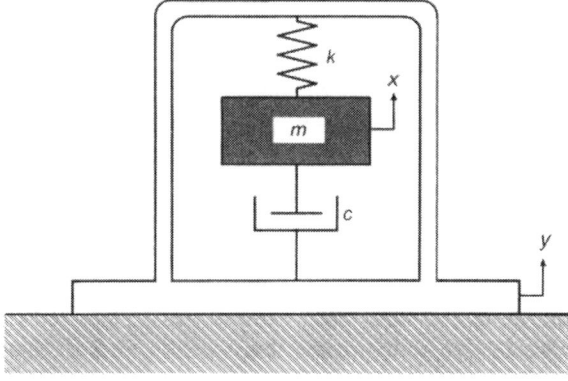

Fig. 4.1 Vibration sensor.

where x and y are the displacement of the seismic mass and the vibrating body respectively. If the relative displacement between the seismic mass and the case is:

$$z = (x - y), \tag{4.5}$$

and if the vibrating body to which the case is attached undergoes sinusoidal motion of the form $y = A \sin \omega t$, then eqn (4.4) becomes:

$$m\ddot{z} + c\dot{z} + kz = m\omega^2 A \sin \omega t. \tag{4.6}$$

This is the well-known equation for the response of a single degree of freedom system to forced vibration. The steady state solution can be assumed to be of the form $z = Z \sin(\omega t - \phi)$ where Z is the amplitude of the mass-case displacement and ϕ is the phase of the displacement with respect to the exciting force. By comparison with standard solutions we can write down eqns (4.7) and (4.8):

$$z = \frac{m\omega^2 A}{\sqrt{(k - m\omega^2)^2 + (c\omega)^2}} = \frac{A(\omega/\omega_n)^2}{\sqrt{[1 - (\omega/\omega_n)^2]^2 + [2\zeta \omega/\omega_n]^2}} \tag{4.7}$$

$$\tan \phi = \frac{\omega c}{k - m\omega^2} = \frac{2\zeta \omega/\omega_n}{1 - (\omega/\omega_n)^2}. \tag{4.8}$$

It is obvious from these expressions that the important parameters are the frequency ratio, ω/ω_n, and the damping factor ζ. Figure 4.2 shows a plot of these equations. The type of sensor (seismometer or accelerometer) is determined by the relationship between the frequency to be measured and the resonance frequency of the transducer.

4.2.1 Seismometers

When the natural frequency of the sensor ω_n is low compared to the vibration frequency being measured, ω, the ratio ω/ω_n is large. The amplitude of the relative displacement Z approaches that of the vibration A, regardless of the value of the damping ratio ζ. In these circumstances the mass m remains stationary while the surrounding case moves with the vibrating body. Sensors of this type are called seismometers. The relative motion z is usually converted to a voltage by making the seismic mass a permanent magnet which moves relative to coils fixed to the case. Since the voltage generated is proportional to the *rate* at which the coils cut the magnetic field lines, the output of the sensor is proportional to the vibration velocity rather than displacement. A typical seismometer will have a resonance frequency between 1 and 5 Hz, and a useful bandwidth from 10 Hz to around 1 kHz.

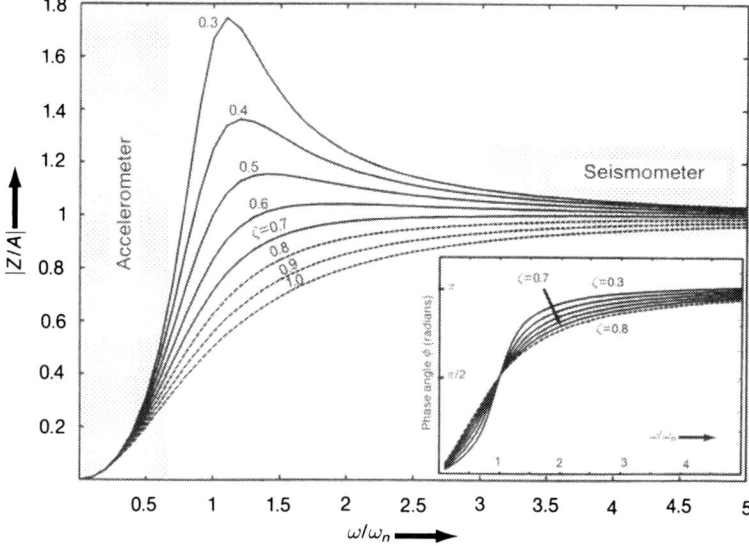

Fig. 4.2 Plot of eqns 4.7 and 4.8.

The main disadvantage of seismometers as vibration sensors is their large size. Since $z = A$, the relative motion of the seismic mass has to be of the same order as that of the vibration being measured and the sensor housing must be large enough to accommodate this motion. For this reason their use in engineering is unusual, and is largely restricted to civil engineering applications.

4.2.2 Accelerometers

When ω_n is high compared to ω the sensor output is proportional to acceleration. Examination of eqn (4.7) shows that the factor:

$$\sqrt{\left[1 - \left(\frac{\omega}{\omega_n}\right)^2\right]^2 + \left[2\zeta\frac{\omega}{\omega_n}\right]^2}$$

approaches unity as ω/ω_n tend towards 0, so that:

$$Z = \frac{\omega^2 A}{\omega_n^2} = \frac{(\text{acceleration})}{\omega_n^2}. \tag{4.9}$$

Thus, in an accelerometer operating well below the transducer resonance, Z (the displacement amplitude of the seismic mass with respect to the case) is proportional to the acceleration of the motion being measured, with a

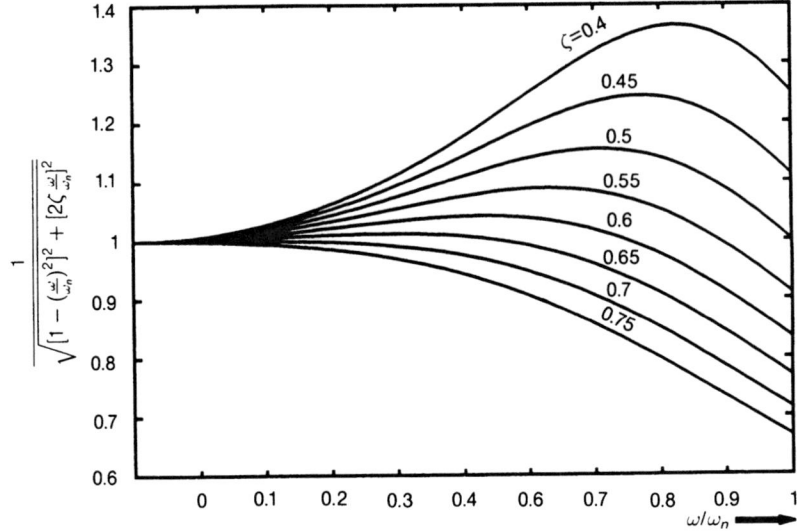

Fig. 4.3 Accelerometer error vs, frequency (ζ as parameter).

factor $1/\omega_n^2$. The useful range of the accelerometer can be seen from Fig. 4.3, which is a plot of:

$$\frac{1}{\sqrt{[1-(\omega/\omega_n)^2]^2+[2\zeta\omega/\omega_n]^2}}$$

for various values of damping ζ. The diagram shows that the useful range of frequencies for an undamped accelerometer is severely limited. However, when $\zeta=0.7$, the useful frequency range extends to around 20% of the resonance frequency, and within this range the maximum error is less than 0.01%.

4.2.3 Phase distortion in accelerometers

The time delay between applying a mechanical input to an accelerometer and the appearance of the resulting electrical output is known as the *phase shift*. If this delay is not the same for all frequencies contained in the mechanical input, the phase relationship between the frequency components of the vibration waveform will be altered, and the resulting electrical output will become distorted. This effect is known as *phase distortion*, and to avoid it we either require the delay to be zero, or else that all frequency components are delayed by the same amount. The first case, a zero delay or zero phase shift, corresponds to $\zeta=0$ and $\omega/\omega_n<1$ (see eqn (4.9)). However, as we have just seen, zero damping is undesirable in an accelerometer.

The second case, an equal timewise phase shift applied to all frequency components, is almost satisfied when $\zeta=0.7$ and $\omega/\omega_n<1$. It can be seen from Fig. 4.2 that when $\zeta=0.7$ and $\omega/\omega_n<1$, the phase angle ϕ is given approximately by:

$$\phi=\frac{\pi}{2}\frac{\omega}{\omega_n}.$$

Thus for $\zeta=0.7$ (and $\zeta=0$) phase distortion is almost eliminated.

4.2.4 Accelerometer resonance frequencies

The resonance frequency of an accelerometer is not constant, although a constant value will be specified on the sensor calibration chart. It depends not only on the seismic mass and the stiffness of the piezoelectric or other transducers to which it is attached, but also on the mass and stiffness of the vibration test object to which the device is attached, and to some extent on the stiffness of the mounting method used.

The situation is illustrated in Fig. 4.4. A seismic mass m_s rests on a transducer, such as a piezoelectric element, which is attached to the transducer base. K is the equivalent stiffness of the transducer and its connection to the base of the device. The mass of the base and housing is m_b. When the accelerometer is not coupled to any other object, as shown in Fig. 4.4(a), the resonance frequency f_r is:

$$f_r=f_s\sqrt{1+\frac{m_s}{m_b}}, \tag{4.10}$$

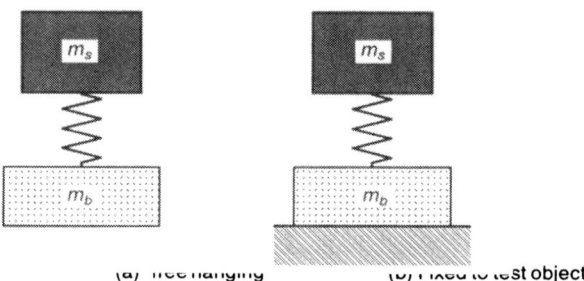

(a) free hanging   (b) Fixed to test object

Fig. 4.4.

where f_s is the resonance frequency of the seismic mass m_s on the stiffness K. The expression for f_s is:

$$f_s = \frac{1}{2\pi}\sqrt{\frac{K}{m_s}}. \tag{4.11}$$

From eqn (4.10) it can be seen that the 'free-hanging' resonance frequency depends upon the ratio of m_s to m_b. At first sight therefore eqn (4.10) seems to imply that the base of the accelerometer should be made as light as possible (i.e. m_b small), so that f_r is as large as possible and the usable bandwidth extended. However, when the accelerometer is mounted on a test object of large mass and stiffness as shown in Fig. 4.4(b), m_b tends to infinity in any case, and the accelerometer resonance approaches f_s. There is therefore no point in trying to make the sensor base light. Other considerations, such as the susceptibility to base strain discussed later, indicate the use of a stiff, and consequently heavy, base.

4.3 Longitudinal velocity sensing

The usual arrangement in a moving coil velocity sensor is shown in Fig. 4.5, where a coil is suspended within a cylindrical magnet. If the coil moves with velocity $\mathrm{d}u/\mathrm{d}t$, the output voltage is:

$$V_{out} = BL \cdot \frac{\mathrm{d}u}{\mathrm{d}t}, \tag{4.12}$$

where B is the strength of the magnetic field in tesla and L the length of conductor cut by the flux.

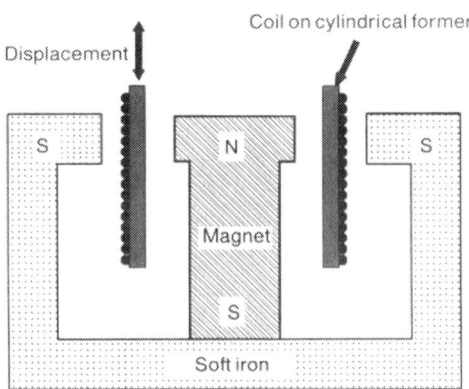

Fig. 4.5 Moving-coil velocity sensor.

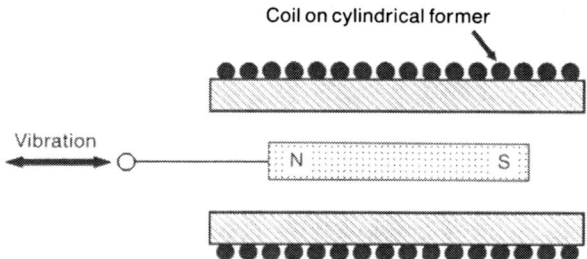

Fig. 4.6 Moving-magnet velocity transducer.

A simple form of linear moving magnet transducer is shown in Fig. 4.6, and consists of a movable magnet within a fixed coil. The output from this type of sensor can be markedly nonlinear however.

4.4 Rotational velocity sensing

The tachometer is the rotational equivalent of the moving coil velocity sensor. In the DC tachometer a magnetic field is generated by a permanent magnet, and a coil is rotated between the poles. The induced DC voltage is proportional to velocity, and is obtained from the rotating coil by means of *slip rings*. Slip rings are highly polished metal disks or cylinders, which allow signals to be transmitted between the rotating and fixed parts of a machine by means of a sliding contact. To ensure that the output voltage is reasonably smooth a number of magnetic poles are usually placed around the coil. However, as the number of poles must be finite, some ripple is unavoidable. In addition, DC tachometers can suffer from noise due to imperfections in the electrical contact at the slip rings.

AC tachometers are in general less noisy than the DC type, since slip rings are unnecessary, and they do not suffer from ripple effects. An AC tachometer consists of a solid rotating conducting cylinder, arranged orthogonally with a pair of coils as shown in Fig. 4.7. One coil is sinusoidally excited at constant frequency, and due to eddy current effects within the rotating core an AC voltage is induced in the second coil. The magnitude of the voltage in the second coil is proportional to the rotation rate.

4.5 Accelerometer designs

An accelerometer is an electromechanical transducer which generates an electrical output when subjected to mechanical shock or vibration.

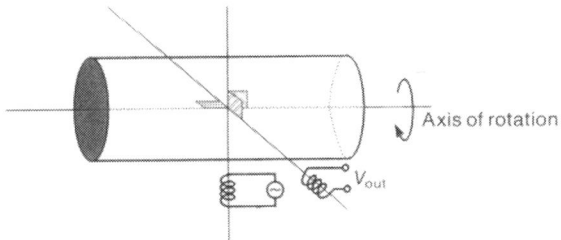

Fig. 4.7 AC Tachometer.

Accelerometers are inertial sensors which make measurements by virtue of Newton's second law. Unlike displacement and velocity, which are usually determined with respect to an arbitrary reference point, acceleration can be measured on an absolute basis.

Shock and vibration measurements are vital to the development, testing and operation of structures and machines in all fields of engineering. Accelerometers are widely used because of their accuracy, robustness, wide frequency response, and sensitivity. In general accelerometers are smaller, lighter, and easier to install than other types of vibration sensor.

In this section the design characteristics, operating principles, and limitations of the most common forms of accelerometer are described. The underlying mathematics are discussed in section 4.2. Clearly, few if any users will wish to design their own accelerometer. However, understanding how a sensor operates is an essential prerequisite to its intelligent selection and use.

4.5.1 Piezoelectric accelerometers

A piezoelectric accelerometer contains two elements; the mass, on which the acceleration acts to produce a force, and a piezoelectric transducer, which converts the force into electric charge.

4.5.1.1 The piezoelectric effect

A piezoelectric material, usually crystalline or ceramic, is one in which a potential difference appears across opposite faces as a result of dimensional changes due to the application of a mechanical force. The effect is reversible, in that a potential difference applied across opposite faces of a block of piezoelectric (PE) material will change its physical dimensions. The PE effect is only possible when the molecular arrangement results in an asymmetric charge distribution as shown in Fig. 4.8. It can be seen

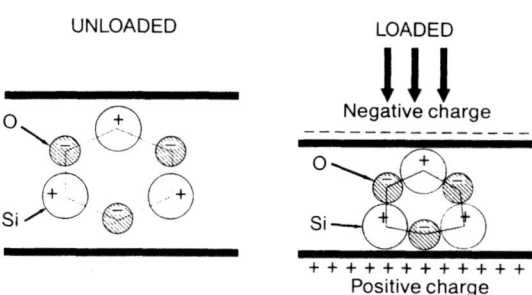

Fig. 4.8 The longitudinal piezoelectric effect in quartz.

that the application of a force F sets up an electrostatic charge on the surfaces. Alternatively, the application of a voltage to the surfaces of the material will change its dimensions. Both longitudinal and transverse PE effects are possible, depending on the molecular geometry. Only the longitudinal effect is illustrated here.

Naturally occurring PE materials include single crystals of quartz and Rochelle salt. However, the PE effect is small in these substances. PE accelerometers normally use man-made PE compounds known as ferroelectric ceramics, which have a much higher responsivity. These consist of mixtures of barium titanate, lead zirconate, and lead metaniobate. Ferroelectric ceramics have the further advantage that, unlike crystalline materials, they can be produced in any desired shape or size.

4.5.1.2 Piezoelectric accelerometer designs

In one of the most common designs (of which an example is shown in Fig. 4.9(a)), the seismic mass rests on a number of piezoelectric disks. The mass and disks are preloaded by a spring and the whole assembly is sealed inside a housing. When the accelerometer is vibrated along its axis the mass exerts a force on the PE material, which develops a variable charge in proportion to the applied force. For frequencies below that of the resonance of the transducer the acceleration of the mass is equal to that of the whole accelerometer.

Three different mechanical constructions are used for PE accelerometers. The most common is the *centre-mounted compression* (CM) type, an example of which has already been shown in Fig. 4.9(a). This construction gives moderate sensitivity and can withstand high levels of continuous vibration or shock without damage. In a CM accelerometer the piezoelectric mass–spring system is mounted on a cylindrical centre post attached to the base of the accelerometer. This design partly isolates the transducer from the surrounding case of the accelerometer, which increases the resonance frequency and thus the usable bandwidth.

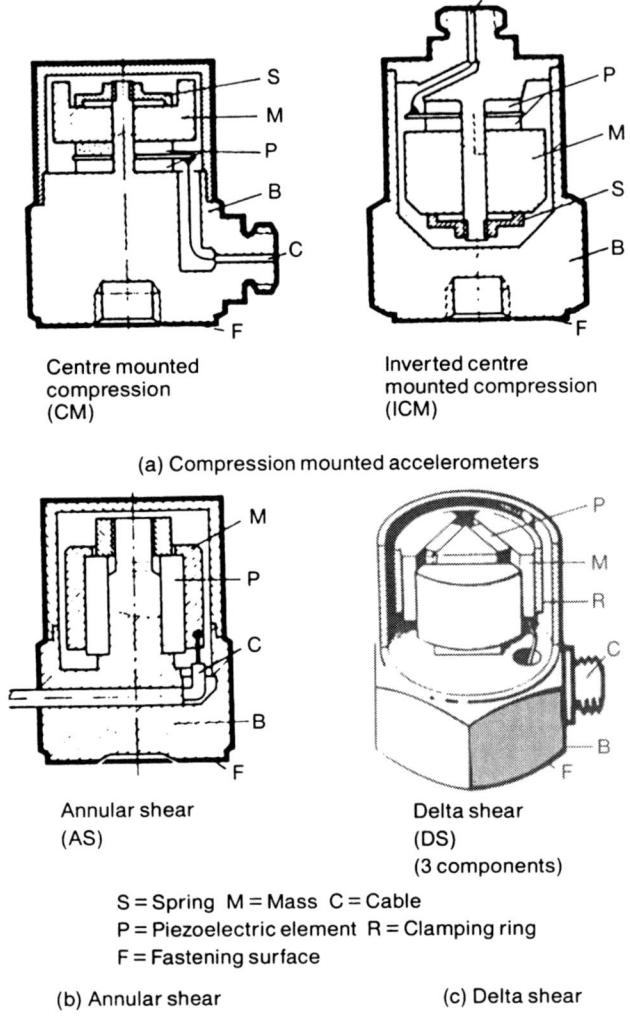

Centre mounted
compression
(CM)

Inverted centre
mounted compression
(ICM)

(a) Compression mounted accelerometers

Annular shear
(AS)

Delta shear
(DS)
(3 components)

S = Spring M = Mass C = Cable
P = Piezoelectric element R = Clamping ring
F = Fastening surface

(b) Annular shear   (c) Delta shear

Fig. 4.9 Single-axis piezoelectric accelerometers (courtesy of Brüel & Kjaer).

However, despite the use of a thick base, CM accelerometers are more sensitive to base strain and thermal transients than other PE designs.

Since no bonding is used in the construction of a CM accelerometer, this type can be used at temperatures up to 250 °C without damage.

The *delta shear* (DS) type have high sensitivity and a high resonance frequency, and are more immune to base strain and thermal effects than

CM sensors. A DS piezoelectric accelerometer contains three piezoelectric elements, arranged in an equilateral triangle as shown in Fig. 4.9(c). Each piezoelectric transducer carries its own seismic mass. Acceleration along the axis of the sensor creates shear stresses in each of the PE elements. The use of shear rather than compression increases the stiffness of the PE elements, which raises the resonance frequency, and makes the accelerometer more immune to external environmental changes than the CM type. However, the more complex construction means that DS piezoelectric accelerometers are usually more expensive than the simpler CM type.

If the seismic masses in a DS accelerometer are bonded to the piezoelectric transducers, the upper operating temperature is limited by the need to avoid softening the bonding adhesive. To avoid this problem the seismic masses are often clamped into position by a preloading ring as shown in Fig 4.9(c). This raises the upper operating temperature to around 250 °C.

The *annular shear* (AS) type represents an attempt to combine the advantages of the DS accelerometer with the simplicity of manufacture and consequent low cost of CM devices. In an AS accelerometer an annular mass is bonded to a piezoelectric ring, which is mounted on a cylindrical centre post as shown in Fig 4.9(b). This approach lends itself to miniaturization and enables very small, light accelerometers with very high resonance frequencies to be constructed. However, the bonding technique used to attach the mass to the piezoelectric element normally limits the maximum upper operating temperature range to around 100 °C.

4.5.2 Frequency response of PE accelerometers

The useful bandwidth of a piezoelectric accelerometer normally extends from a few Hz to about 1/3rd of the resonance frequency. Thus PE accelerometers are used mainly for sensing frequencies in excess of 5–10 Hz.

The low frequency cut-off obtained when using a PE accelerometer depends mainly on the electronic amplifier to which the sensor output is connected, rather than being a feature of the sensor itself. The low frequency limit is determined by the RC time constant formed by the accelerometer output impedance and the preamplifier input impedance, and is usually less than 5 Hz. To obtain lower frequencies preamplifiers with very high input impedances ($>10^9 \Omega$) must be used.

4.5.3 Cross-axis (transverse) sensitivity of PE accelerometers

An ideal accelerometer should be sensitive to motion along a single axis, and should not produce an output when it is accelerated in any direction at 90° to this axis. Real accelerometers approach this ideal with varying degrees of success. However, there is always some sensitivity to vibration perpendicular to the intended sensing axis. The size of the cross-axis sensitivity depends upon the exact orientation of the sensor with respect to the motion, but is usually within the range 0–5%. The dependence of transverse sensitivity on orientation is a consequence of the fact that the maximum charge and voltage sensitivity of the PE transducer elements is often not perfectly aligned with the axis of the accelerometer. This produces directions of maximum and minimum transverse sensitivity, which are at 90° to each other and perpendicular to the main axis. The calibration details supplied with an accelerometer always specify the maximum transverse sensitivity. The direction of minimum sensitivity is often marked with a spot on the accelerometer housing.

We have already seen that an accelerometer has a resonant response to vibration along its main axis. Similarly a resonance will occur if it is exposed to vibration in transverse directions. In the region of this resonance the transverse sensitivity may reach 100% of the main axis (off-resonance) sensitivity. For most PE accelerometers the transverse resonance frequency occurs at about 30% of the main-axis resonance frequency, and therefore lies within the bandwidth normally considered as the useful operating range of an accelerometer. Thus, spurious results may be generated if an accelerometer is subjected to transverse vibration at around 1/3 of the main resonance frequency, or if it is exposed to transverse shock loads which can contain broad-band energy extending to high frequencies.

PE accelerometers should not be dropped or knocked, since the transverse loads so produced may exceed the design limits and can cause irreparable damage.

4.5.4 Piezoresistive accelerometers

The piezoresistive effect occurs in silicon and other materials, and is used in the fabrication of miniature accelerometers. As the name implies, a change in electrical resistance occurs in response to changes in the applied stress. Piezoresistive (PR) vibration sensors are formed by placing stress-sensitive resistors on highly stressed parts of a suitable mechanical structure. The PR transducers are usually attached to cantilevers, or other beam configurations, and are connected in a Wheatstone bridge circuit. The beam may carry a seismic mass or may utilize its own self-weight. Under acceleration the beam deflects due to the inertial forces and undergoes stress changes. These stress variations are converted into an electrical output, which is proportional to acceleration, by the PR transducers.

PR accelerometers are relatively easy to construct, provide a low frequency response extending to DC, and work well over a relatively large temperature range (−50 to +150 °C). A further feature which makes them valuable is their ability to include signal processing and communication functions within the sensor package at little extra cost.

The drawbacks of PR devices are that the output signal level is moderate (typically 100 mV full scale for a 10 V bridge excitation), the sensitivity can be temperature dependent, and the usable bandwidth is not as large as that which may be obtained from a PE sensor.

Before considering the design of PR accelerometers in detail, we need to understand the phenomenon of piezoresistance.

4.5.4.1 Analysis of piezoresistance

If a rectilinear resistor has length l, width w, thickness t and a bulk resistivity ρ, its resistance R will be:

$$R = \frac{\rho l}{wt}. \tag{4.13}$$

The gauge factor or strain sensitivity is defined as k, where:

$$k = \frac{\mathrm{d}R/R}{\varepsilon} \tag{4.14}$$

ε is the relative change in length of the resistor (the strain) due to a stress, σ, applied to the substrate parallel to its length. Figure 4.10 shows the consequences of the applied stress. The length increases by an amount $\mathrm{d}l$, while the width and thickness decrease by $\mathrm{d}w$ and $\mathrm{d}t$ due to Poisson's ratio v. It is clear that $\mathrm{d}w = -vw\varepsilon$ and $\mathrm{d}t = -vt\varepsilon$.

The original cross-section was:

$$A = wt.$$

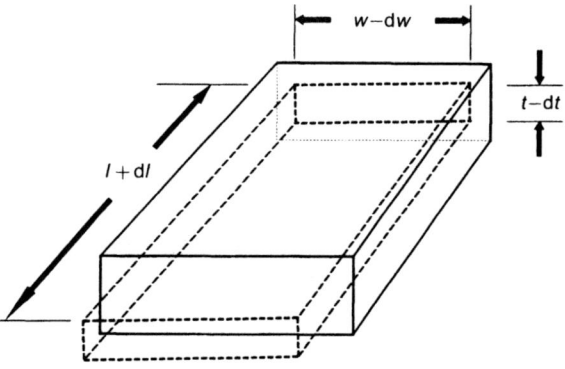

Fig. 4.10 Geometrical changes due to applied stress.

Owing to the strain ε, the new cross-sectional area is:

$$A' = (w - dw)(t - dt)$$

$$= wt + 2vwt\varepsilon + v^2 wt\varepsilon.$$

The term $(v^2 wt\varepsilon)$ is very small compared with the other two terms in the equation and can be neglected. We can therefore write the change in cross-sectional area as:

$$A - A' = dA = -2v\varepsilon A,$$

giving:

$$\frac{dA}{A} = -2v\varepsilon.$$

Differentiating eqn (4.13) gives:

$$\frac{dR}{R} = \frac{d\rho}{\rho} + \frac{dl}{l} - \frac{dA}{A},$$

hence the gauge factor k is:

$$k = \frac{d\rho/\rho}{\varepsilon} + (1 + 2v). \tag{4.15}$$

Typically v will be between 0.2 and 0.3. Equation (4.15) therefore shows that the longitudinal gauge factor is a function of changes in both longitudinal resistivity and geometry. In a conventional foil or wire strain gauges the piezoresistive effects are negligible, and the variations in resistance are mainly a function of dimensional changes. For a foil gauge k is approximately 2. For PR strain gauges the first term in eqn (4.15) is significant, and higher gauge factors (typically around 10) can be achieved, giving enhanced sensitivity. It should be noted however that the resistivity of most PR materials is strongly temperature dependent, and that as a result PR strain gauges generally have a higher thermal sensitivity than other types.

4.5.5 Silicon piezoresistive (PR) accelerometers

One of the first silicon accelerometers was demonstrated in 1976 [1]. It consisted of a single cantilever carrying PR strain gauges near its root. The device was fragile and required the inclusion of a liquid-filled cell for damping. Improved designs have since appeared.

Three types of silicon PR accelerometer have since evolved, as shown in Fig. 4.11. These are the *single cantilever*, the *doubly supported* structure, and the *'top hat'* design. It will be noted that, despite its name, the single cantilever design can have one, two or more supporting beams carrying

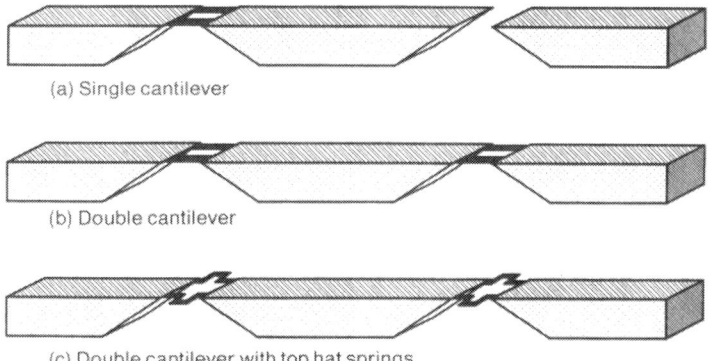

(a) Single cantilever

(b) Double cantilever

(c) Double cantilever with top hat springs

Fig. 4.11.

the seismic mass. The distinguishing feature of this type is that the support beams are all placed along one edge. The doubly supported cantilever of Fig. 4.11(b) uses four supporting beams, and for this reason is sometimes referred to as a quad cantilever. The top hat approach is implemented when a device capable of undergoing large displacements is required. If the supports are folded as shown in Fig. 4.11(c) the effective length can be increased, while the package size is kept constant.

In all three types viscous damping is provided by the inclusion of a small volume of air.

The single cantilever type has the highest gain, but can also have a large transverse sensitivity. The doubly supported and top hat designs provide good off-axis cancellation and are reasonably robust. They are consequently the most common configuration for PR silicon accelerometers.

4.5.5.1 Resonance frequency

The resonance frequency of a silicon PR accelerometer is determined by the stiffness of the support structure and the seismic mass as discussed previously. Typical resonance frequencies for silicon PR accelerometers are in the range 500–5000 Hz. Thus the bandwidth is considerably lower than that of a typical PE device. For comparison, thick film devices (see later) usually have resonance frequencies between 300 and 2000 Hz.

4.5.5.2 Sensitivity of silicon accelerometers

The sensitivity increases with seismic mass m_s, decreases with support stiffness K, and is modulated by a transduction efficiency term b:

$$\text{sensitivity} = b \cdot \frac{m_s}{K}.$$

The main parameters which determine b are the position of the PR transducers, the number of PR transducers, and the transduction efficiency of the transducers, which is a function of their geometry and chemical constitution.

The sensitivity is also inversely proportional to the square of the resonance frequency f_r:

$$\text{sensitivity} = \frac{(2\pi)^2 b}{f_r^2}. \tag{4.16}$$

Equation (4.16) shows that if a sensitive accelerometer is required the resonance frequency should be as low as possible. Thus the sensitivity requirement is in conflict with the need for a large bandwidth. A typical doubly supported design with an output of 5 mV/g per volt of bridge excitation will have a resonance around 500 Hz, implying the useful bandwidth extends from DC to around 150 Hz. If the design is modified so that the resonance moves to 37 kHz, which is approaching the practical limit for this type of accelerometer, the sensitivity decreases to 5 μV/g per volt of bridge excitation.

4.5.5.3 Off-axis modes and transverse sensitivity

A major advantage of the doubly supported type is a substantial reduction in transverse sensitivity and unwanted resonant modes compared with the single cantilever. The three principal modes of vibration for a doubly supported cantilever are shown in Fig. 4.12. The intended axis of sensitivity is vertical, and therefore only the mode shown in Fig. 4.12(a) is desired. However, for the modes shown in Figs 4.12(b) and (c), which are

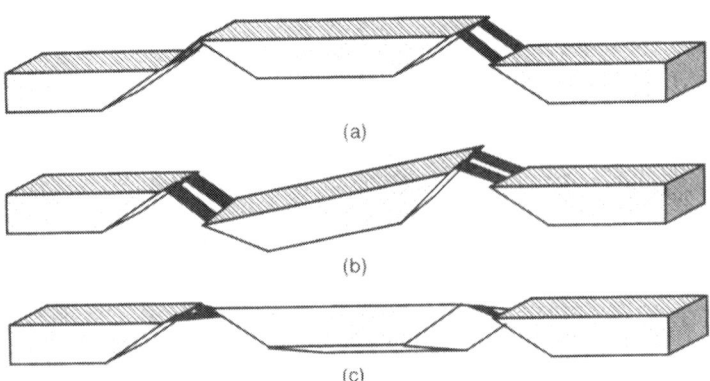

(a)

(b)

(c)

Fig. 4.12 Three principal modes of vibration for doubly-supported cantilever.

excited by off-axis acceleration, opposite sides of the structure undergo opposite forms of bending. Careful positioning of the PR transducers and the use of a bridge circuit can therefore be used to give a high degree of immunity to outputs originating from these modes. No such symmetry exists in the case of a single cantilever. The maximum transverse sensitivity of a doubly supported PR silicon accelerometer is comparable with that of a PE device, and is typically 5% of the main axis sensitivity.

4.5.6 Thick film piezoresistive accelerometers

Thick film circuits are formed by the deposition of layers of special pastes (usually referred to as inks, although there is little resemblance to conventional ink) on to an insulating substrate. The printed pattern is fired in a manner akin to the production of pottery, to produce electrical pathways of a controlled resistance. Parts of a thick film circuit can be made sensitive to strain or temperature. The thick film pattern can include mounting positions for the insertion of conventional silicon devices, in which case the assembly is known as a *thick film hybrid*. The process is relatively cheap, especially if large numbers of devices are produced, and the use of hybrid construction allows the sensor housing to include sophisticated signal conditioning circuits. These factors indicate that thick film technology is likely to play an increasingly important part in automotive sensor design.

Three main categories of thick film inks exist: conductors, dielectrics (insulators), and resistors. Conductors are used for interconnections, such as the wiring of a bridge circuit. Dielectrics are used for coating conducting surfaces (such as steel) prior to laying down thick film patterns, for constructing thick film capacitors, and for insulating cross-over points, where one conducting path traverses another. Resistor inks are the most interesting from the point of view of sensor design, since many thick film materials are markedly piezoresistive.

The main constituents of a thick film ink are the *binder* (a glass frit), the *vehicle* (an organic solvent), and the *active elements* (metallic alloys or oxides). After printing, each layer of a thick film pattern is dried to remove the organic solvents (the vehicle) which give the ink its viscosity. Drying also improves the adhesion properties, bonding the ink to its substrate and rendering the pattern immune to smudging. This stage is usually performed in a conventional oven at 100–150 °C.

A final high-temperature firing is required to remove any remaining solvent and to sinter the binder and the active elements. During the firing cycle a thick film pattern is raised to a temperature between 500 and 1000 °C. The glass frit melts, wets the substrate, and forms a continuous matrix which holds the functional elements. The heating and cooling gradients, the peak temperature, and the dwell time determine the *firing profile*. This has a critical effect on the production of a thick film circuit, since it allows the electrical characteristics of the inks to be modified. Resistor materials are especially sensitive to the firing profile, and the resistor layer is usually therefore the last to be fired. However, the need for passivation of a circuit often necessitates covering it with a dielectric layer. To avoid changing the resistor values a low-melting-point dielectric is often used for the final layer.

The need for high-temperature firing can cause problems if thick film piezoresistors are to be applied to previously heat-treated components. The temperatures used can adversely affect, for example, the properties of toughened or hardened steels.

Thick film circuits and sensors are created by screen printing. This is essentially a stencil process, in which the printing ink is forced through the open areas of a mesh-reinforced screen on to the surface of a substrate. The screen stencils are formed by photolithography. In this process a photosensitive mesh-filling material is exposed to ultra-violet light through a mask depicting the required pattern. The image is photographically developed, and those parts of the pattern which have not been fixed are subsequently washed away.

The use of thick film technology was introduced as a means of miniaturizing circuits without incurring the expense associated with fabrication in silicon. It was soon noted that thick film materials had temperature and stress-dependent properties. Although this was awkward from the point of view of circuit fabrication, it has since been turned to good account in sensor design. The linear temperature coefficient of resistance (TCR) possessed by certain platinum-containing conductive inks has allowed resistance thermometers to be constructed wholly in thick film form [2]. More importantly from the point of view of accelerometer design, the PR properties of thick film resistor (TFR) inks can be used to form strain sensors. This approach has been used to make a number of pressure sensors [see for example ref. 3], and is currently being exploited to produce accelerometers [4].

4.5.7 Capacitive accelerometers

The use of a capacitance change is well-established in sensor design as a measuring technique. Pressure and displacement have frequently been evaluated by this means, and a number of manufacturers now supply capacitive accelerometers. These are claimed to give better performance than conventional types for measuring low frequency, low level acceleration.

Capacitive accelerometers are micromachined from single crystal silicon. A typical design is shown in Fig. 4.13. A conducting layer is deposited

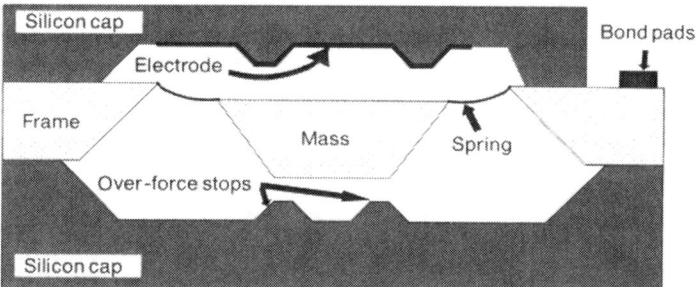

Fig. 4.13 Cross-section through typical silicon capacitive accelerometer.

on to one surface of a silicon block. A second conducting layer is laid down on one side of a second block, which acts as the seismic mass. The mass is supported by beams, and is separated from the base by an air gap as shown in Fig. 4.13. The two halves of the sensor are electrostatically bonded together. Signal processing electronics can be incorporated within the sensor package.

Capacitive accelerometers are expensive to manufacture, and usually cost about the same as the PE equivalent. They are claimed to provide a frequency response down to DC, stable damping characteristics, and a useful bandwidth which is larger than PR but smaller than that provided by PE devices. In addition their output is extremely stable over a wide temperature range. As shown by Table 4.1, this is their primary advantage over the other, more common accelerometer types.

The output impedance of any capacitive sensor is intrinsically very high. However, if a capacitive accelerometer includes signal conditioning circuits within the package a low output impedance is usually provided, so that vibration signals may be transmitted over lengthy cables without loss or distortion.

To sum up therefore, the inherent advantages of capacitive sensing are its low power, high output, wide dynamic range, and relative immunity to thermal effects. Unfortunately the signal processing requirements for this type of sensor are not straightforward, and it can be difficult to obtain a linear output.

4.5.8 Environmental effects on accelerometers

Accelerometers are frequently used to make measurements under severe environmental conditions. These can include both high and low temperatures, severe shock loadings up to several hundred g, a wide range of

Table 4.1
Comparison of different accelerometer types

Parameter	Piezoelectric	Piezoresistive Silicon	Thick film	Capacitive
DC response	No	Yes	Yes	Yes
Bandwidth	Wide	Moderate	Low	Wide
Self-generating	Yes	No	No	No
Impedance	High	Low	Low	V. high
Signal level	High	Low	Low	Moderate
Temperature range (°C)	−55 to 100*	−55 to 150	−50 to 120	−200 to 200
Linearity	Good	Moderate	Moderate	Excellent
Static calibration (turnover)	No	Yes	Yes	Yes
Cost	High	Low	Low	High
Ruggedness	Good	Moderate	Moderate	Good
Suitable for shock	Yes	No	No	No

*Unless special designs are used.

humidities (up to 100%), exposure to electromagnetic radiation, and exposure to potentially damaging chemicals such as petrol, oil, hydraulic fluid, battery acid and water. It is therefore important that accelerometers are sealed, are as rugged as possible, and have a sensitivity to environmental changes which is as low as possible.

4.5.9 Thermal sensitivity of accelerometers

Table 4.1 shows the temperature ranges over which the various types of accelerometer may be used with confidence. Piezoelectric accelerometers often begin to give erroneous results above 100 °C, and should not be used at high temperatures unless the manufacturer states this is permissible.

In addition to the temperature limits shown in Table 4.1, accelerometers can also exhibit a slowly varying output when subjected to temperature transients. This output arises from two causes:

- a pyroelectric effect; and
- non-uniform thermal expansion of the accelerometer structure, which subjects the sensing element to a stress variation.

The effect of a temperature transient is usually seen as a low-frequency electrical output, which can be removed by appropriate filtering. It is unlikely to interfere with measurements unless very low-frequency, low amplitude vibrations are being investigated.

4.5.10 Humidity

Most commercial accelerometers are of sealed construction, with either welded or epoxy bonded housings. They have a high resistance to the majority of corrosive agents encountered in industry. If it is necessary to immerse an accelerometer, or to use it in an environment where heavy condensation is likely, it may be necessary to take special precautions with cabling. Impervious Teflon cables can be obtained for this purpose. These should be sealed at the accelerometer/cable connection using a silicone rubber sealant. This material is suitable for use from -70 to $+260\,°C$.

4.5.11 Acoustic sensitivity of accelerometers

All accelerometers are to some extent sensitive to acoustic excitation. However, with careful design unwanted acoustically-generated outputs can be avoided. Most accelerometers are of rigid, mechanically isolated construction, and pressure variations in air will have little effect on the force transmitted to the sensing element. In general, vibrations of the surface to which the sensor is attached will give rise to much higher signals than will direct acoustical excitation of the accelerometer. Care must be taken when trying to measure very low level accelerations in the presence of an intense sound field.

4.5.12 Base strain sensitivity

When an accelerometer is mounted on the surface of a structure undergoing strain variations, a spurious output will be generated by the accelerometer if any of the strain is transmitted to the sensing element. The *base strain sensitivity* is usually specified by the manufacturer for individual accelerometers, in units of acceleration (g or $m \cdot s^{-2}$) per microstrain (μ).

Commercial accelerometers usually have thick, stiff bases to reduce strain effects. A further reduction in base strain sensitivity is possible for PE accelerometers by using delta shear designs.

4.5.13 Electromagnetic interference (EMI) and accelerometers

Most commercial accelerometers are housed within welded steel or stainless steel enclosures. These normally provide adequate screening from the effects of electric or magnetic fields, which arise from spark ignition systems and other electrical wiring. However, EMI problems can arise in the cabling used to connect accelerometers to a measurement system. The problem is particularly common when piezoelectric (PE) sensors are used, since these devices have a very high impedance. If PE transducers are used in an automotive application it is desirable to employ the type which incorporates a charge amplifier within the sensor housing. The sensor output is then low-impedance, and is consequently much less likely to be affected by EMI.

4.5.14 Accelerometer mounting techniques

It is very important to mount an accelerometer properly. If it is not rigidly fixed to a vibrating surface the high-frequency performance will be poor. The mounting technique commonly recommended for commercial sensors uses a steel stud, and is shown in Fig. 4.14. If this technique is used it is essential that a good surface contact is obtained between the accelerometer and the test specimen. The mounting surface should therefore be smooth and flat, with the hole for the mounting stud perpendicular to the surface. A thin film of silicon grease can be applied to the base of the accelerometer before screwing it down, to improve the mounting stiffness and hence the high-frequency performance.

4.5.15 Accelerometer connecting cables

Piezoelectric and capacitive sensors are very high impedance devices. They are therefore susceptible to triboelectric noise, which can be generated by connecting cables when these are subjected to mechanical motion. Dynamic bending, compression, or tension of cables momentarily separates the cable screen from the dielectric, and this results in local capacitance changes and the release of triboelectric charge as shown in

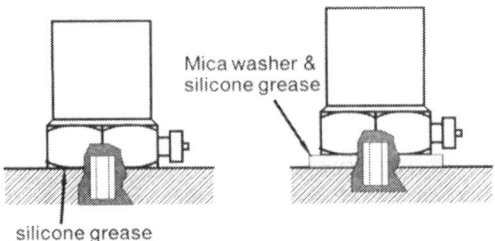

Fig. 4.14 Accelerometer mounting techniques.

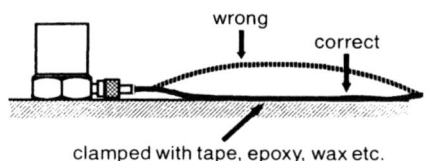

Fig. 4.15 Accelerometer cable clamping.

Fig. 4.15. The problem is especially severe at low frequencies, and when using conventional coaxial cable rather than specially-treated accelerometer leads. To minimize, this source of electrical noise accelerometer cables having special noise reduction treatment should be used. These should be clamped to the vibrating specimen using epoxy glue, wax or adhesive tape so that relative movement is avoided as far as possible.

In some circumstances it may not be possible to clamp the cable to the vibrating surface. For example, it may be too hot. In this case an accelerometer with a top-mounted output connector should be used, and the cable should be led away from the accelerometer and clamped as soon as possible.

References

[1] A batch-fabricated silicon accelerometer. L.M. Roylance and J.B. Angell, in *IEEE Transactions on Electron Devices*, Volume ED-26 no. 12, page 1911, 1979.
[2] Thick-film platinum temperature sensors. Q.M. Reynolds and M.G. Norton, in *Proceedings of the Test and Transducer Conference*, 1985, Wembley, London, Volume 2 pages 31–44.
[3] Thick film resistor strain gauges: New applications. B. Morten, M. Prudenziati and A. Taroni, *Proc. International Kongress Mikroelektronik*, pages 345–54, November 1982.
[4] *A novel accelerometer using thick-film technology*. R. Sion, J. Atkinson, J. Turner. Sensors & Actuators A, volumes 37–8, pages 348–51, 1993.

Strain measurement techniques   5

5.1 Introduction

The strain in a material is the amount of mechanical deformation suffered by that material due to the action of a force. Strain is a vector quantity, in that it has both magnitude and direction, and it is measured using *strain gauges*, which should be orientated in the direction of measurement. Strain determination is essentially a measurement of displacement.

Early methods of strain measurement were mechanically based, and used systems of levers to magnify the small extensions due to a strain until they were large enough to move a pointer against a calibrated scale. Mechanical strain gauges are still sometimes used, their great advantage being that they require no external power supply. Gauges with a pointer indication are only useful for static strain measurements. For use with dynamic loads scratch gauges can be used, in which the system of magnifying levers actuates a stylus which 'writes' a trace onto a revolving foil or glass disk actuated by clockwork. However, they are generally larger than the much more common electrical resistance strain gauge. Since strain is a point property, the use of finite-sized gauges gives an error arising from the fact that the strain measurement is integrated over the gauge length.

The electrical resistance strain gauge (which from now on we shall simply refer to as a strain gauge) was developed in the 1930s. Early versions simply consisted of a length of wire glued to the test object. Changes in length (strains) on the surface of the test object are transferred to the wire, and cause alterations in the resistance of the wire. The resistance changes can be very accurately measured by means of a bridge circuit (see later). Modern foil gauges are often made by etching a thin metal foil, rather than using a wire, but the principle of operation is identical in each case. In recent years semiconductor and thick film strain gauges have become available, both of which are many times more sensitive than wire or foil gauges.

5.2 Wire and foil strain gauges

To ensure that the resistance changes are as large as possible a long gauge length must be used. However, it is also necessary to make the gauge

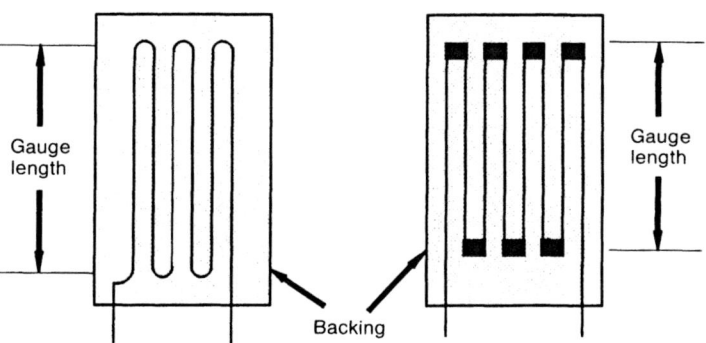

Fig. 5.1 (a) Wire, (b) Foil strain gauges.

occupy as small an area as possible, so that the measurement approximates to point strain determination. To achieve these two objectives the conductor (foil or wire) is normally folded as shown in Fig. 5.1. The etching process used to manufacture foil gauges can be made to produce complex shapes for special purposes, such as the examples shown in Fig. 5.2.

The change in resistance of a gauge is related to the change in gauge length (the strain) by the *gauge factor k*:

$$k = \frac{\delta R/R}{\delta L/L} = \frac{\delta R}{\varepsilon R},$$

(5.1)

where: R = original gauge resistance (i.e. without strain),
δR = change in gauge resistance,
L = gauge length,
δL = change in gauge length,
ε = strain.

The conductor used to manufacture a strain gauge should have as high a gauge factor as possible, so that small strains cause large changes in resistance. In addition, the gauge factor must be linear—in other words, multiples of a given extension must produce the same multiple of the resistance change. The gauge factor for a wire or foil gauge is usually about 2. The exact value for a given gauge is supplied with the gauge, and is normally quoted to two decimal places. Table 5.1 gives the characteristics of typical commercially available wire and foil gauges.

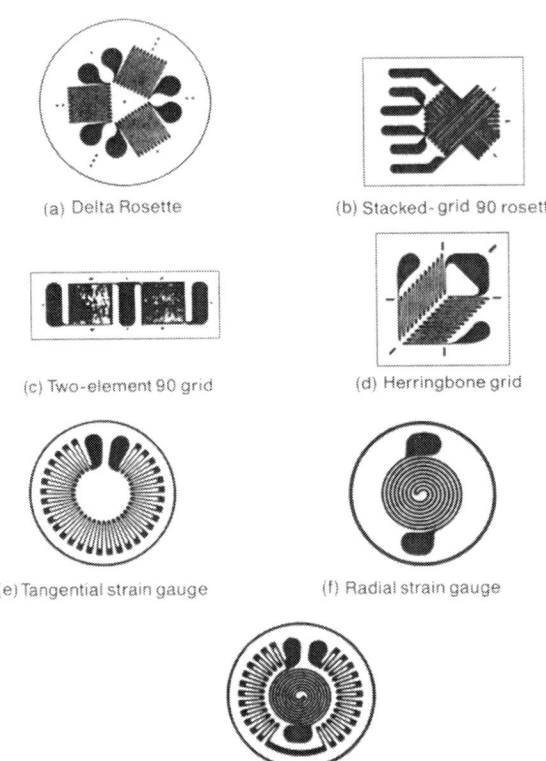

(a) Delta Rosette

(b) Stacked-grid 90 rosette

(c) Two-element 90 grid

(d) Herringbone grid

(e) Tangential strain gauge

(f) Radial strain gauge

(g) Gauge for measuring radial and tangential strain

Fig. 5.2 Example strain gauge configurations.

Table 5.1
Characteristics of typical foil and wire strain gauges

Gauge factor	Approximately 2. Exact value supplied with gauge
Gauge resistance	Standardised values: $120\,\Omega$, $350\,\Omega$, $600\,\Omega$ and $1000\,\Omega$ are used
Linearity	Usually within $\pm 0.1\%$ up to $4000\,\mu\varepsilon$, and within $\pm 1\%$ up to $10{,}000\,\mu\varepsilon$ (N.B. $\mu\varepsilon$ denotes microstrain: $1\,\mu\varepsilon \equiv 0.0001\%$ strain).
Breaking strain	About $25{,}000\,\mu\varepsilon$
Fatigue life	Up to 10 million strain reversals
Temperature compensation	Gauges may be obtained with coefficients of thermal expansion that match general purpose steels, stainless steels, and aluminium alloys. Compensation for thermal effects may also be obtained by the use of bridge circuits.

5.3 Semiconductor strain gauges

Semiconductor strain gauges consist of a strip of semiconducting material such as silicon or germanium, doped with a controlled amount of impurity to give the desired characteristic. The gauge factor may be as high as 50–60, making it possible to measure extremely small strains. However, semiconductor strain gauges are often very sensitive to temperature variations, and are generally far less rugged than foil or wire gauges. Table 5.2 gives typical values for commercial semiconductor strain gauges.

5.4 Thick film strain gauges

These devices were described in some detail in chapter 4. Thick film strain transducers are more sensitive than foil types, but are not as sensitive as semiconductor versions. The values given in Table 5.3 are typical.

5.5 Strain gauge transducers

Strain gauges form the active sensing element in a number of transducers, notably load cells and pressure sensors.

Table 5.2
Typical characteristics of semiconductor strain gauges

Gauge factor	50–60. Exact value supplied with gauge
Gauge resistance	$> 500\,\Omega$
Linearity	Within $\pm 1\%$ up to $1000\,\mu\varepsilon$
Breaking strain	About $5000\,\mu\varepsilon$
Fatigue life	Up to 1 million strain reversals

Table 5.3
Typical characteristics of thick film strain transducers

Gauge factor	10–20
Gauge resistance	$> 10\,k\Omega$
Linearity	Within $\pm 1\%$ up to $1000\,\mu\varepsilon$
Breaking strain	About $5000\,\mu\varepsilon$
Fatigue life	Up to 10 million strain reversals

A typical force transducer sensor is shown in Fig. 5.3, and consists of a ring which is deformed into an oval shape under the action of a load. When the loading is as shown on Fig. 5.3 the strain gauges on the inside of the ring experience tension, while those on the outside undergo compression. The four gauges are connected in a bridge circuit, and the sensitivity of the arrangement is four times that which can be achieved with a single gauge.

Another common form of load cell consists of a cylinder which is compressed by the load. As shown in Fig. 5.4 four strain gauges are

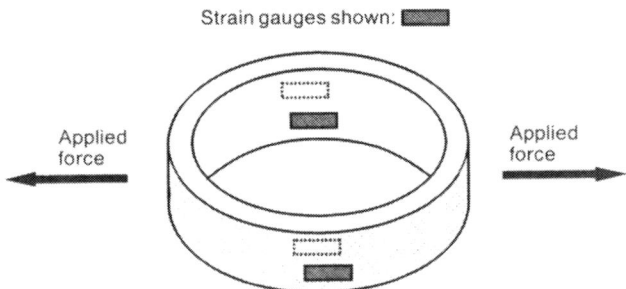

Fig. 5.3 "Proof ring" load cell with strain gauges.

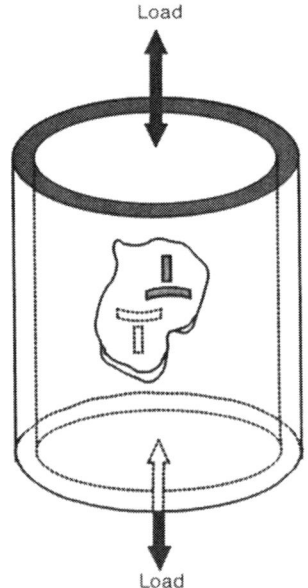

Fig. 5.4 Cylindrical load cell and strain gauges.

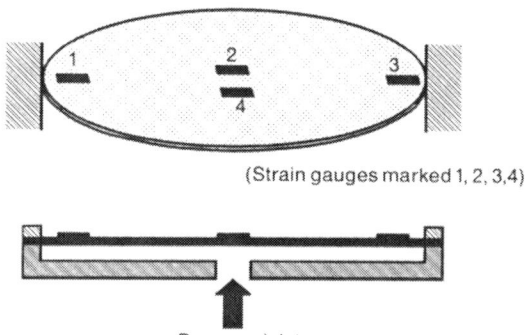

(Strain gauges marked 1, 2, 3, 4)

Pressure inlet

Fig. 5.5 Diaphragm and strain gauge pressure transducer.

attached to the cylinder, usually on the inside so that the gauges are protected against damage by abrasion. Two of the gauges are placed so that they measure compression in the cylinder walls, and two so that they measure the resulting lateral deformation due to Poisson's ratio. Poisson's ratio is around 0.3 for most engineering metals, so this arrangement is 2.6 times more sensitive than a single gauge as explained later (see eqn (5.11)).

A common form of pressure transducer consists of a thin circular diaphragm which is deformed by the action of pressure. A pair of strain gauges (one on each side) are mounted at the centre of the diaphragm, and the remaining pair are mounted in the comparatively unstressed area at the edge of the diaphragm as shown in Fig. 5.5. The unstressed gauges are known as dummy gauges, and are included in the bridge circuit to provide automatic temperature compensation (as shown in section 5.6.3). The gauges at the centre of the diaphragm give an output which is proportional to pressure. Further details of strain-based pressure transducers are given in chapter 6.

5.6 Bridge circuits for strain gauge transducers

As discussed in chapter 1, many modulating sensors control the flow of electrical energy by means of changes in resistance or impedance. These changes can be quite small: for example, the change in resistance of a $120\,\Omega$ foil strain gauge at $1000\,\mu\varepsilon$ is typically of the order of $0.001\,\Omega$. Bridge circuits provide the most convenient way of measuring these small changes.

Bridges may be excited by an AC or DC power supply. DC bridges are always composed of resistances, and are referred to as resistive bridges. An AC bridge can contain resistors, inductors or capacitances, and is known as an impedance bridge.

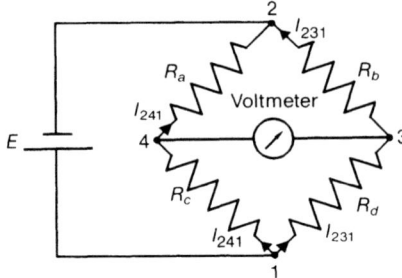

Fig. 5.6 Wheatstone (resistor) bridge.

The best-known example of a resistive bridge is the Wheatstone bridge, an example of which is shown in Fig 5.6. A bridge simply consists of a pair of parallel potential dividers which are connected across an excitation voltage E. When all four resistances have an equal value R the bridge is said to be *balanced*, i.e. a voltmeter connected across the centre of the circuit will register zero.

Considering the voltages across each resistor in turn, and using notation based on Fig. 5.6, we can write:

$$V_{24} = I_{241} R_a = V_{23} = I_{231} R_b \quad \text{and} \quad V_{14} = I_{241} R_c = V_{13} = I_{231} R_d,$$

so

$$I_{241} R_a = I_{231} R_b, \tag{5.2}$$

and

$$I_{241} R_c = I_{231} R_d. \tag{5.3}$$

Dividing (5.2) by (5.3) gives:

$$\frac{R_a}{R_c} = \frac{R_b}{R_d}. \tag{5.4}$$

Equation (5.4) shows that a change in the resistances on one side of the bridge can be balanced by adjusting resistance values on the other side of the bridge. It will be demonstrated later that this forms the basis of a method for compensating strain gauge bridge circuits, so that ambient temperature changes do not affect the measurement. Furthermore, eqn (5.4) suggests that the sensitivity of a bridge can be enhanced if more than one sensor is included in the measuring circuit.

5.6.1 Bridge balancing

If one or more of the resistances in a DC bridge are sensors, some means for *balancing* the bridge (i.e. ensuring that the voltage V in Fig. 5.6 is

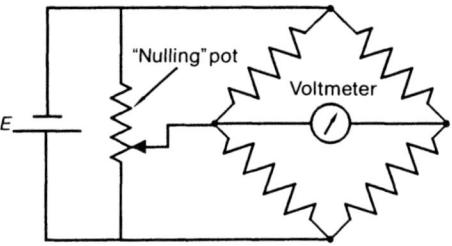

Fig. 5.7 Method of balancing a Wheatstone bridge.

zero) before making a measurement must be provided. If strain is to be measured, the normal sequence of events is to balance the bridge, then apply a load to the test object, and finally to calculate the strain from the out-of-balance voltage produced by the load. The simplest method of balancing a bridge is to make one of the resistors a potentiometer. However, this prevents the construction of bridge circuits in which all four arms are sensors. Figure 5.7 shows an alternative way of balancing a resistor bridge. The advantage of this method is that it can be used regardless of how many fixed resistors and sensors make up the bridge.

We can see from Fig. 5.6 that a bridge may contain up to four sensors. Bridges with one sensor (and three fixed resistors) are referred to as quarter bridges. Bridges with two sensors are known as half bridges, and bridges with four sensing elements are called full bridges. The behaviour of each configuration is analyzed in the following sections.

5.6.2 The quarter bridge

If only one of the resistances shown in Fig. 5.6 is a sensor and the other three are fixed resistances, a change in the parameter being sensed will alter the bridge output V. Suppose R_a is a strain gauge with an initial resistance R. When a strain is applied the value of R_a becomes $R + \delta R$, while the other resistances remain unchanged (and equal to R). Then:

$$I_{241} = \frac{E}{R_a + R_c} = \frac{E}{2R + \delta R}$$

$$V_{24} = R_a I_{241} = (R + \delta R)\left(\frac{E}{2R + \delta R}\right)$$

$$= \left(\frac{ER + E\delta R}{2R + \delta R}\right).$$

If we consider the other side of the bridge:

$$V_{23} = R_b I_{231} = R_b \left(\frac{E}{R_b + R_d} \right) = \frac{E}{2}$$

(since $R_b = R_d = R$). So the out-of-balance voltage V across the bridge is:

$$V = V_{23} - V_{24} = \left(\frac{E}{2} \right) - \left(\frac{ER + E\delta R}{2R + \delta R} \right)$$

$$= \frac{2ER + E\delta R - 2ER - 2E\delta R}{4R + 2\delta R}.$$

If δR is small, we can say that $4R + 2\delta R \approx 4R$. Thus, since δR may have either sign we can write:

$$V = \pm \frac{E\delta R}{4R}. \tag{5.5}$$

Substituting eqn (5.1) into this result, we find that if R_a is a strain gauge, the strain ε is given by:

$$\varepsilon = \frac{4V}{Ek}, \tag{5.6}$$

where k is the gauge factor. Equation (5.6) shows that if the excitation voltage E and the gauge factor k are known for a quarter bridge, the strain may be determined by measuring the out-of-balance voltage.

If a strain gauge is used with a coefficient of thermal expansion which differs from that of the object being measured, it is impossible to say whether a measured strain is due to an applied load, differential thermal expansion, or a mixture of the two. However, an important feature of bridge circuits is that compensation for thermal effects may be obtained by incorporating a compensating resistor into one of the arms of the bridge. This resistor must have the same temperature characteristic as the sensor, and it is common to use a dummy sensor as compensating resistor. With strain gauges for example, it is usual to place a gauge identical to that used for sensing on an adjacent, unstressed part of the test object. This dummy gauge is then included in the bridge circuit, and compensates for the effect of any temperature changes on the active or sensing gauge.

The compensating resistor in a quarter bridge is connected in a position adjacent to the sensor. Thus, if for example R_a in Fig. 5.6 is a strain gauge, the dummy gauge must be R_b or R_c.

Suppose R_a is an active strain gauge, and R_c is a dummy gauge for temperature compensation. Let the effect of a temperature change be to increase the resistance of both R_a and R_c by an amount δR. Then:

$$I_{241} = \frac{E}{R_a + R_c} = \frac{E}{2R + 2\Delta R}$$

$$V_{24} = R_a I_{241} = (R + \Delta R)\left(\frac{E}{2R + 2\Delta R}\right)$$

that is:

$$V_{24} = \frac{E}{2}.$$

As R_b and R_d are unaffected by the change in temperature, V_{23} is unaltered. The out-of-balance voltage V is:

$$V = V_{23} - V_{24} = \frac{E}{2} - \frac{E}{2} = 0. \tag{5.7}$$

The effect of any temperature-induced resistance change has been cancelled out, and the bridge is still balanced in the absence of strain. If the active gauge is subjected to a strain and a temperature change while the dummy gauge is subjected to a temperature change alone, the thermal effects cancel out. Any out-of-balance voltage measured across the bridge with this arrangement is due to strain alone.

5.6.3 The half bridge

Figure 5.8 shows a bridge circuit containing two sensors. A bridge containing a pair of sensors in adjacent positions has two main advantages over the single-sensor arrangement discussed in the last section. First, the sensitivity is enhanced, because the additional sensor gives a larger out-of-balance voltage. Second, the thermal changes to each sensor cancel out, as shown in the previous section.

If we consider as an example the strain-gauged cantilever beam undergoing simple bending shown in Fig. 5.9, we see that under the applied load the top surface of the beam experiences tensile strain, and the lower surface (numerically equal) compression. The resistance of strain gauge R_a becomes $R + \delta R$ under the action of the bending, while that of R_c becomes $R - \delta R$. We may calculate the out-of-balance voltage expected with this arrangement as follows:

$$I_{241} = \frac{E}{R_a + R_c} = \frac{R}{2R}$$

$$V_{24} = R_a I_{241} = (R + \delta R) \cdot \left(\frac{E}{2R}\right) = \frac{ER + E\delta R}{2R}.$$

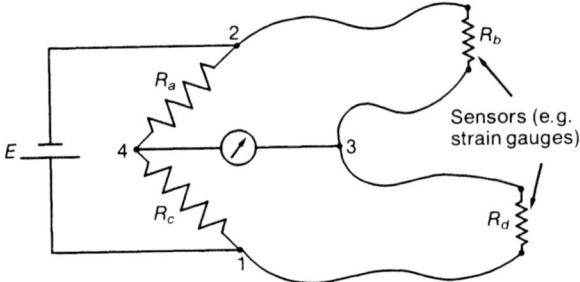

Fig. 5.8 The half-bridge.

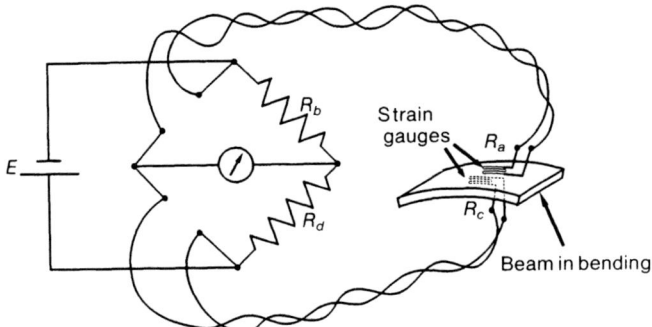

Fig. 5.9 Strain gauge half-bridge used to measure bending.

We saw earlier that $V_{23} = E/2$, so the out-of-balance voltage V is:

$$V = V_{23} - V_{24} = \left(\frac{E}{2}\right) - \left(\frac{ER + E\delta R}{2R}\right) = \pm\frac{E\delta R}{2R}. \qquad (5.8)$$

Comparing eqn (5.8) with eqn (5.5), we see that a half bridge has twice the sensitivity of a quarter bridge.

If the cantilever has a uniform cross-section, its neutral axis will coincide with the centre of the beam. The stress at any point within the beam can then be calculated from the measured surface strains.

If the cantilever is subjected to a tensile force in addition to bending, as shown in Fig. 5.10, both strain gauges will be stretched by the same amount. The form of the resulting stress distribution is sketched in Fig. 5.10. Since the active gauges are connected in adjacent arms of the measuring bridge, the extension of each gauge due to the tensile load will cancel out in the same way as a thermal gauge extension. This can be useful if only the bending component is to be measured, but obviously

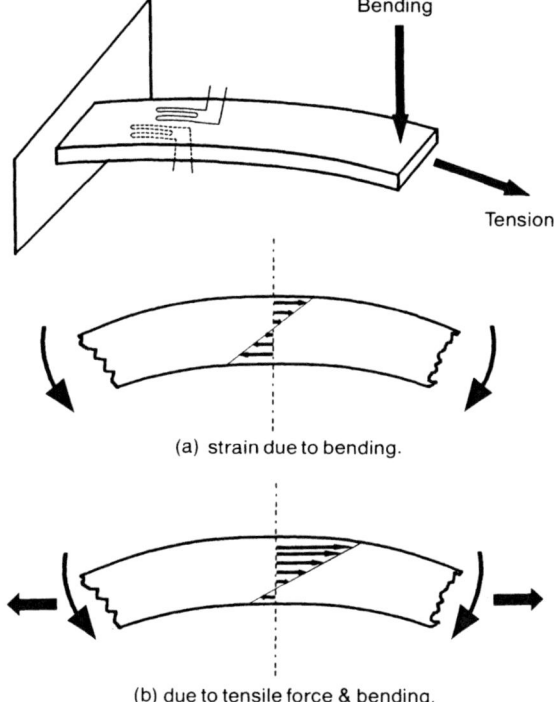

(a) strain due to bending.

(b) due to tensile force & bending.

Fig. 5.10 Strain-gauged beam undergoing tension and bending, with strain distributions.

any calculation of total stress made from such a strain measurement will be in error by the amount of the tensile stress.

Figure 5.11 shows a tensile test, in which a pair of orthogonally mounted strain gauges have been applied to the test specimen. With the gauges connected as shown in Fig. 5.11 the effect of the strain is to increase the resistance of R_a to $R+\delta R$. If an object is subjected to longitudinal tensile stress it will extend in the direction of the stress, and contract in the transverse direction. The transverse strain is proportional to the logitudinal strain, and the constant of proportionality is known as Poisson's ratio v. For most metals v is in the range 0.28–0.32, with 0.3 being a typical value.

If the test specimen of Fig. 5.11 has a Poisson's ratio v, the effect of the tensile stress will be to decrease the value of R_c to $R-v\delta R$. The out-of-balance voltage in this case can now be calculated:

$$I_{241} = \frac{E}{R_a + R_c} = \frac{E}{2R+(1-v)\delta R}$$

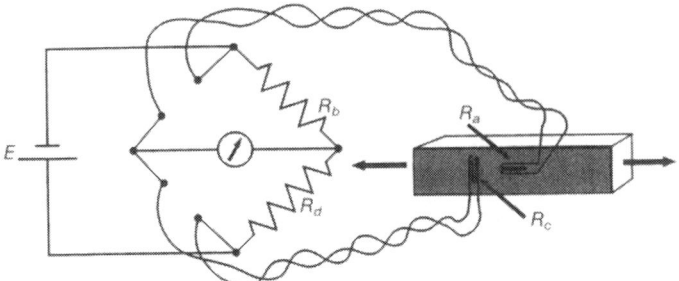

Fig. 5.11 Strain gauge half-bridge used to measure tensile strain.

$$V_{24} = R_a I_{241} = \left(\frac{ER + E\delta R}{2R + (1-v)\delta R} \right).$$

Once again $V_{23} = E/2$, so the out-of-balance voltage V is:

$$V = \frac{\pm(1+v)E\delta R}{4R + 2(1-v)\delta R}.$$

If δR is small compared to R, $4R + 2(1-v)\delta R \approx 4R$ so that:

$$V \approx \pm \frac{(1+v)E\delta R}{4R}. \tag{5.9}$$

Comparing eqns (5.9) and (5.5), we see that the sensitivity of a half bridge used in this way to measure tensile strain is 1.3 times that of a quarter bridge for a material where $v = 0.3$.

In addition to improving the sensitivity by 30%, the arrangement described above also provides automatic temperature compensation. As described earlier, if temperature changes occur during a test and the gauges are not matched to the material under investigation, both gauges will experience the same change in resistance, and the thermal effects will cancel out.

5.6.4 The full bridge

The foregoing discussion of quarter and half bridges has shown that it is possible to enhance the sensitivity obtained from a measuring bridge if more than one active sensor is connected into the circuit. The greatest possible sensitivity is obtained when all four arms of a bridge contain active sensors. When a full bridge is used it is not usually possible to adjust the resistance of one of the arms of the bridge to balance it, and an arrangement such as that shown in Fig. 5.7 must be used.

With a full strain gauge bridge all four gauges are usually exposed to the same temperature variations, and automatic thermal compensation is obtained.

Figure 5.12 again shows a strain gauged cantilever undergoing bending. When the gauges are connected as shown the following resistance changes occur under the load:

R_a increases to $R + \delta R$
R_b decreases to $R - \delta R$
R_c decreases to $R - \delta R$
R_d increases to $R + \delta R$.

Calculating the out-of-balance voltage as before it is simple to show that the out-of-balance voltage V is:

$$V = \pm \frac{E \delta R}{R}. \qquad (5.10)$$

By comparing eqns (5.10) and (5.5) we see that a full bridge used to measure bending strain has four times the sensitivity of a bridge containing only one strain gauge.

The use of a full bridge in tensile strain measurement gives greater sensitivity than a quarter or half bridge. Figure 5.13 shows a tensile test specimen which has been instrumented with four strain gauges, two mounted longitudinally and two in the transverse direction. If the test material has a Poisson's ratio v, the following changes in gauge resistance

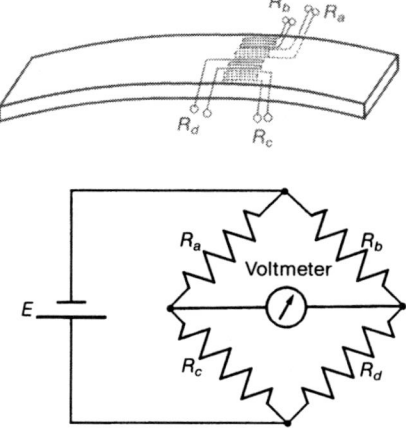

Fig. 5.12 Bending strain measurement, full bridge.

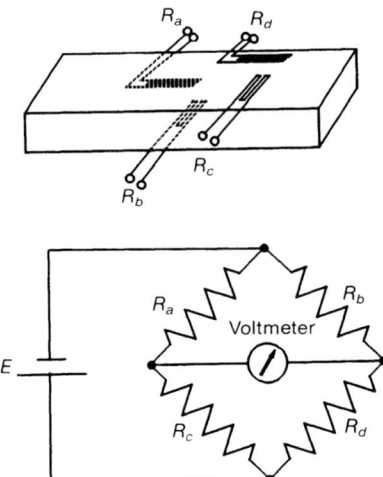

Fig. 5.13 Bending strain measurement, full bridge.

will take place under load:

$$R_a \text{ increases to } R+\delta R$$
$$R_b \text{ decreases to } R-\nu\delta R$$
$$R_c \text{ decreases to } R-\nu\delta R$$
$$R_d \text{ increases to } R+\delta R.$$

Proceeding as usual to calculate the out-of-balance voltage V, we find that:

$$V = \pm E\frac{(1+\nu)\delta R}{2R}. \tag{5.11}$$

If eqn (5.11) is compared with eqn (5.5), we see that a full bridge used for tensile measurement gives a sensitivity 2.6 times higher than a single strain gauge for a material with a Poisson's ratio of 0.3.

There are many other methods of using strain gauge bridges for stress measurement, and the examples in this section are not intended to be an exhaustive list. However, the basic principles discussed are common to all measuring bridges, and on the whole other arrangements are 'variations on a theme'.

5.6.5 AC bridges

The majority of variable inductance or capacitance sensors are used as part of an AC bridge. AC bridges containing resistive sensors are also

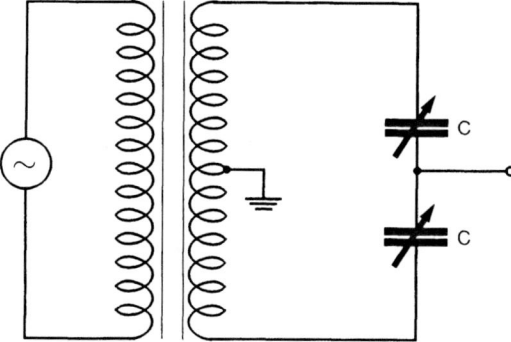

Fig. 5.14 AC bridge (capacitive sensors in this case) excited via a centre-tapped transformer.

common, since unwanted reactive effects due to the capacitance or inductance of connecting leads can be cancelled out.

The simplest form of AC excited impedance bridge is obtained by replacing the DC excitation voltage E in Fig. 5.6 by an oscillator. However, it is more usual to replace one side of such a bridge by a transformer, as shown in Fig. 5.14, since this ensures that the two sides of the bridge are excited in phase opposition with equal magnitude. The transformer either has a centre-tapped secondary or a pair of identical secondary windings. If the centre tap is connected to ground and the other two positions are sensors a single-ended output is obtained.

Special arrangements are used for bridge circuits incorporating push–pull devices. A push–pull transducer is arranged in two parts, such that a change in the sensed parameter produces a decrease in the output of one part and a simultaneous increase in the output of the second part. A number of sensors (for example the LVDT[1]) are available in push–pull form. A push–pull sensor can be connected into a bridge containing two resistors in one of two ways, as shown in Fig. 5.15. At balance an AC bridge must simultaneously satisfy two conditions: the real and the imaginary parts of the out-of-balance voltage must be zero. If for example inductive sensors with a low resistance are used, the contribution to the out-of-balance voltage from changes in transducer resistance is negligible in comparison to the inductance change, and to a first approximation the bridge shown in Fig. 5.15(a) will give an output:

$$V_{\text{out}} = \left(\frac{V_s}{2}\right) \cdot \left(\frac{\Delta L}{L}\right),$$

---

[1] Linear variable differential transformer—see chapter 3.

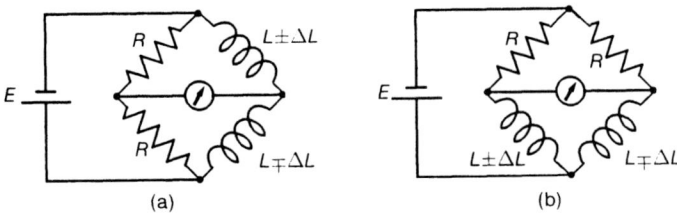

Fig. 5.15 Alternative connections for push–pull sensors: (a) on one side of the bridge (b) both sides of the detector.

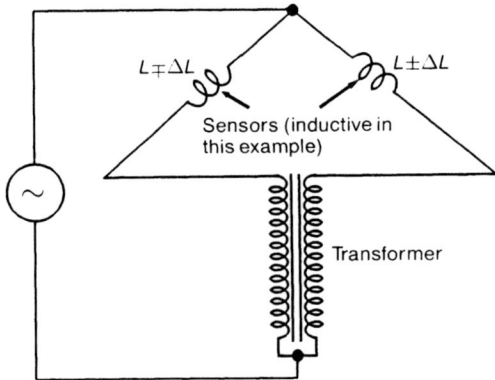

Fig. 5.16 AC transformer bridge (often called a Blumlein bridge).

where V_{out} and V_s are complex (AC) voltages. Under similar conditions, and if $\omega L = R$, the output of the circuit shown in Fig. 5.15(b) will be twice that of Fig. 5.14(a).

An alternative form of AC bridge which is often used with push–pull inductive sensors is shown in Fig. 5.16. Two of the bridge arms are inductively coupled (generally by means of a transformer), and a push–pull sensor makes up the other two arms. This circuit is known as a Blumlein bridge, and it simplifies earthing and shielding problems since stray cable capacitances appear in parallel across the measurement arms and are consequently ignored. In general a Blumlein bridge with two closely coupled inductive arms gives better noise immunity and a greater constancy of sensitivity than an AC bridge with uncoupled inductive or purely resistive arms.

5.7 Summary

The preceding discussion has shown how strain may be measured using quarter, half and full bridges, and how a reasonable degree of immunity from thermal errors may be achieved. It should be stressed that the signal voltages produced by a strain gauge bridge are often very small. For example, in a quarter bridge with an excitation of 10 V, where the active strain gauge experiences 200 µε, the bridge output will be only 1 mV. Care must therefore be taken to screen all signal leads against noise, and to place an amplifier as close to the bridge as possible. Most strain gauge amplifiers are based on operational amplifiers, or op-amps. Suitable circuits are discussed in a later chapter.

The subject of experimental strain measurement is complex, and this chapter only provides an introduction to the subject. For further details refer to the works listed in the references.

References

[1] *Measurement Systems Application and Design*, by E.O. Doebelin. 4th Edition, McGraw-Hill, 1990.
[2] *Strain Measurements*, by J. Vaughan. Bruel and Kjaer, 1975. [Now out of print, but worth obtaining through libraries.]
[3] *Instrumentation Reference Book*, edited by B.E. Noltingk. Butterworths, 1988.
[4] *Theory and Practice of force Measurement*, by A. Bray *et al*. Academic Press, 1990.

6.1 Introduction

When a fluid (liquid or gas) comes into contact with a surface it produces a force perpendicular to the surface. The force per unit area is called the pressure. The SI unit of pressure is the Pascal (Pa). 1 Pascal is equivalent to 1 Newton/m^2. Other units still in use in engineering include pounds per square inch (psi or lb/in^2), atmospheres (atm), millimetres of mercury (mmHg), and bars. Table 6.1 gives conversion factors between the various units of pressure.

Pressure measurements may be divided into three categories, namely *absolute pressure*, *gauge pressure*, and *differential pressure*. The absolute pressure is the difference between the pressure at a particular point in a fluid and the absolute zero of pressure, i.e. a complete vacuum. A mercury barometer is an example of an absolute pressure sensor, since the height of the column of mercury measures the difference between atmospheric pressure and the 'zero' pressure of the Torricellian vacuum that exists above the column of mercury.

If a pressure sensor measures the difference between an unknown pressure and local atmospheric pressure, the measurement is known as *gauge pressure*. The circular dial-type pressure gauges fitted to steam boilers normally indicate gauge pressure. This convention presumably arose because the amount of work obtained from a steam engine is a function of the amount by which the supply steam pressure exceeds that of the atmosphere.

If the pressure transducer measures the difference between two unknown pressures, neither of which is atmospheric, then the measurement is known as *differential pressure*. The difference between absolute, gauge, and

Table 6.1

Conversion factors for units of pressure

1 Pa	= 1 N/m^2	= 1.45×10^{-4} lb/in^2
1 lb/in^2	= 6895 N/m^2	= 0.0703 kg/cm^2
1 atm	= 101,325 N/m^2	= 14.7 lb/in^2
1 bar	= 100,000 N/m^2	= 14.5 lb/in^2
1 mmHg	= 133.3 N/m^2	= 1.93×10^{-2} lb/in^2

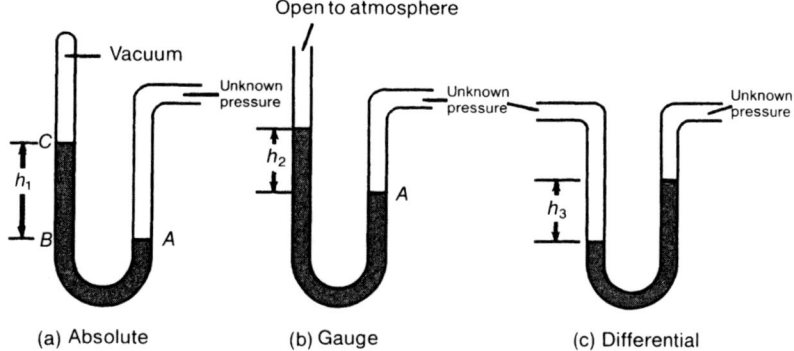

Fig. 6.1 Absolute, gauge and differential pressure measurements.

differential pressure measurement is shown in Fig. 6.1 using mercury manometers.

There are three fundamental means by which a pressure may be measured. The simplest approach involves balancing the unknown pressure against the pressure produced by a column of liquid of known density. Instruments using this principle are called *manometers*. The analysis of a manometer is straightforward. Consider a simple U-tube containing a liquid of density ρ, as shown in in Fig. 6.1. If we consider Fig. 6.1(a), the points A and B are at the same horizontal level when the device is in equilibrium. The liquid at C stands at a height h_1 above B. Then the pressure at A is P_A, where:

$$P_A = P_B \ (P_B \text{ is the pressure at } B)$$
$$= \text{pressure due to column of liquid } BC$$
$$= h_1 \rho.$$

In the case of Fig. 6.1(b), the unknown pressure at A is:

$$P_A = h_2 \rho + (\text{atmospheric pressure}),$$

since the manometer is open to atmospheric pressure at one side, rather than being sealed as was the case in Fig. 6.1(a). The analysis is similar for Fig. 6.1(c), except that in this case the atmospheric pressure has been replaced by a second unknown pressure.

The second method of pressure measurement involves allowing the unknown pressure to act on a known area. The resulting force is measured either directly or indirectly. Devices of this type are called *dead-weight testers*, and they are normally only used for calibrating other forms of pressure sensor. The basic arrangement is shown in Fig. 6.2. It consists

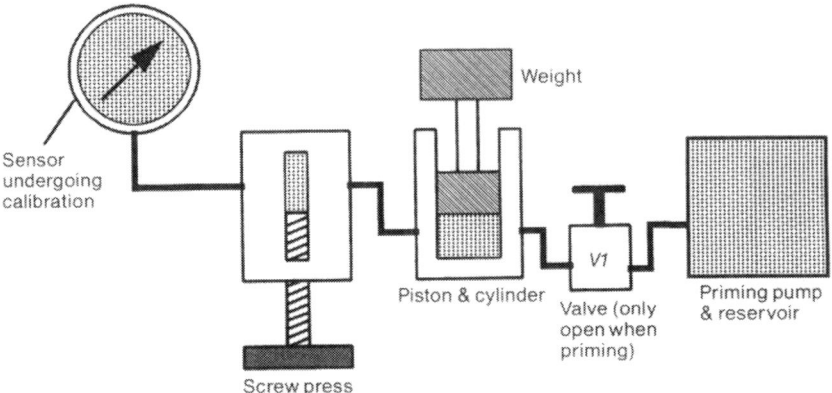

Fig. 6.2 Dead weight pressure measurement system for sensor calibration.

of a priming pump and reservoir, an isolating valve, a weighted piston, a screw press, and the pressure sensor under test. The sequence of events is as follows. First, the screw press is extended to its zero position. Weights representing the desired pressure are applied to the piston. The priming pump is operated to pressurize the system, and the priming valve (*V1* on the diagram) is closed. The screw press is then adjusted until the pressure in the system has increased sufficiently to just raise the piston off its end-stops. Neglecting any friction forces, if the pressure on the piston is P N/m^2, and its area is A m^2, then the resulting force on the piston is PA Newtons, which will support a weight $W = PA$ Newtons. The accuracy of the calibration depends on the precision with which the piston and its associated cylinder are made, and on the degree to which friction has been eliminated from the system.

In the third approach the unknown pressure is allowed to act on an elastic structure of known area and properties. Most commercial pressure sensors adopt this approach. The resulting stress, strain, or deflection is measured in a variety of ways.

The most common form of pressure sensor is probably a diaphragm fitted with strain gauges, although capacitive systems are used almost as frequently. Figure 6.3 shows diaphragm-type pressure sensors configured to measure differential, gauge, and absolute pressure.

6.2 Elastic pressure sensors

In theory a very wide range of elastic structures can be used for pressure measurement. The literature is full of ingenious devices employing novel

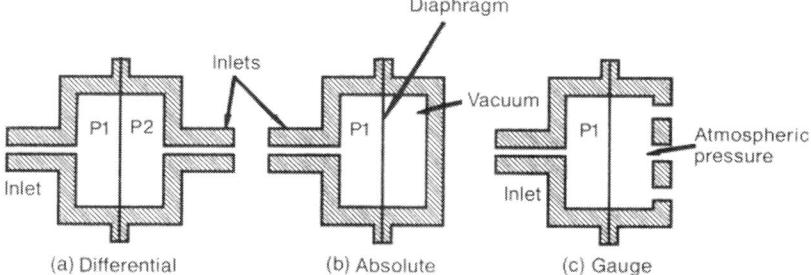

Fig. 6.3 Differential, absolute and gauge pressure sensor (Diaphragm type).

forms of architecture [1]. However, in practice almost all of those which have reached commercial production use one of three approaches, and are based on either diaphragms, bellows, or Bourdon tubes. Diaphragms are probably the most common, although bellows are also used. Bourdon tubes are mainly confined to laboratory measurement systems.

It should be noted that pressure sensors based on an elastic structure have two components; the elastic structure, which forms the primary sensing element and deflects in response to an applied pressure, and a secondary transducer, which senses this deflection and converts it to (usually) an electrical signal. The most common form of secondary transducer is the strain gauge, although piezoelectric and piezoresistive devices, potentiometers, differential transformers, variable capacitors and variable inductors have all been used.

6.2.1 Bourdon tubes

The Bourdon tube is the basis of many mechanical pressure sensors (particularly the familiar circular moving-pointer type). Bourdon tubes are also used in some electrical transducers, particularly those where the output displacement is to be sensed by potentiometers or differential transformers, both of which normally require rotary actuation. The basis of all forms of Bourdon tube is a tube of non-circular (and usually flat-sided or oval) cross-section as shown in Fig. 6.4. One end is sealed, and pressure is applied. The flat sides bulge outward, as though the tube were trying to attain a circular cross-section. The resulting distortion tends to straighten the tube. The end of a C-type Bourdon tube undergoes a curved displacement as shown in Fig. 6.4(a), while the spiral and twisted types produce angular motion. Theoretical analysis of the behaviour of a Bourdon tube is difficult, and usually some form of finite-element analysis is needed. Bourdon tube pressure sensors can be markedly non-linear, and often display an unwanted

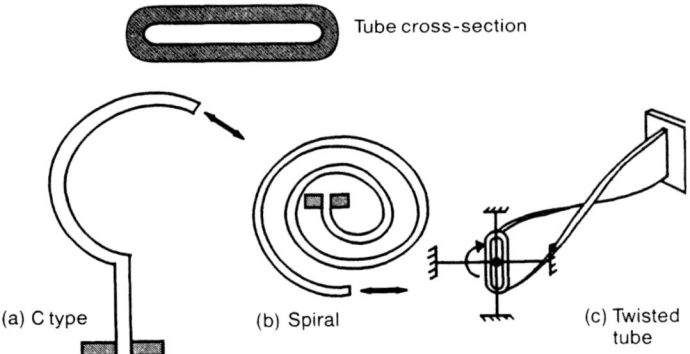

Tube cross-section

(a) C type   (b) Spiral   (c) Twisted tube

Fig. 6.4 Bourdon tubes.

thermal sensitivity. They also suffer from hysteresis errors, which are usually of the order of 1–2% of full scale deflection.

C-type Bourdon tubes have been used for pressures up to about 7×10^8 N/m^2 (100 000 psi). The spiral and twisted versions produce larger displacements, and are mainly used below 7×10^6 N/m^2 (1000 psi). The best accuracy that can be achieved is usually around 0.1%.

The free end of a twisted Bourdon tube is usually supported by an arrangement such as that shown in Fig. 6.4(c), which is stiff in all radial directions but soft in rotation. This helps to protect the device from damage due to shock loads and vibration.

6.2.2 Bellows

Figure 6.5 shows two configurations of bellows. The deflection of a bellows is usually more linear than that of a Bourdon tube. They are reversible with low hysteresis, and are often found in pneumatic systems where they act as pressure/displacement transducers. However, the most common application is undoubtedly in the production of low-cost aneroid barometers for atmospheric pressure measurement.

Bellows are manufactured in a variety of materials. The spring rate (modulus of compression) is proportional to the modulus of elasticity of the material from which the bellows is formed, and to the cube of the wall thickness. It is also inversely proportional to the number of convolutions and to the square of the outside diameter of the bellows [1].

6.2.3 Diaphragms and membranes

Diaphragms are probably the most popular elastic structure used in pressure sensors. They can be subdivided into two types; thin membranes

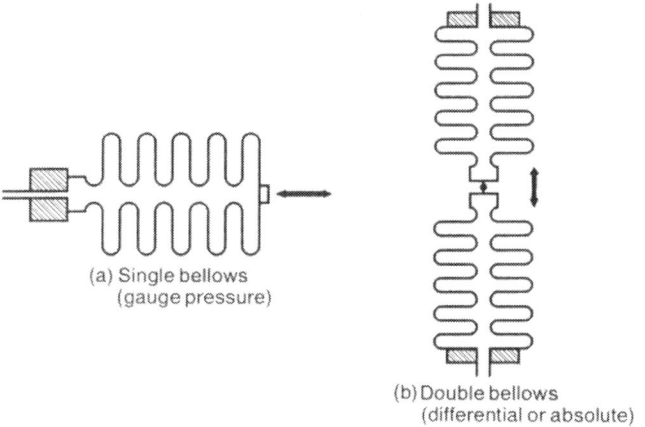

(a) Single bellows
(gauge pressure)

(b) Double bellows
(differential or absolute)

Fig. 6.5 Bellows pressure sensors.

under radial tension, which form part of an inductive or capacitive pressure sensor, and thicker diaphragms or plates, used in conjunction with resistive or piezoelectric transducers. The membrane type are the most sensitive, and can be used for applications such as measuring low-pressure fluctuations. The most familiar example of a membrane pressure sensor is undoubtedly the capacitor microphone. Where larger pressures are to be measured a thin circular plate is used which is strong enough to carry strain gauges. The plate is either clamped around its circumference by a pair of solid rings, or alternatively the whole assembly may be machined from a solid block of material.

In order that the relationship between applied pressure and deflection is reasonably linear, the centre deflection of the plate must not exceed half its thickness. The approximate design equations for a flat circular diaphragm of this form are as follows (see Fig. 6.6 for nomenclature and units):

$$\text{centre deflection } d_m = \frac{3(1-v^2)D^4p}{256\,Yt^3} \qquad (6.1)$$

(d_m is linearly related to pressure p if $d_m \leqslant 0.\,5t$). The maximum circumferential stress s_m is:

$$s_m = \frac{3D^2p}{16t^2}. \qquad (6.2)$$

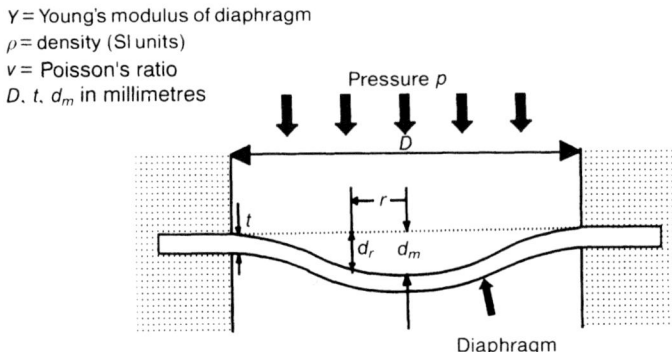

Y = Young's modulus of diaphragm
ρ = density (SI units)
v = Poisson's ratio
D, t, d_m in millimetres

Pressure p

Diaphragm

Fig. 6.6 Flat circular diaphragm pressure sensor.

Since a diaphragm has stiffness and mass it will resonate. The lowest natural frequency (for an air or gas medium) is given by eqn (6.3):

$$f_0 \simeq \frac{10^4 t}{\pi D^4} \sqrt{\frac{Y}{3\rho(1-v^2)}} \ \text{(Hz)}. \qquad (6.3)$$

For example, using mild steel with a Young's modulus of 210 GPa, Poisson's ratio of 0.3, and density 7800 kg/m³, a 10 mm diameter diaphragm 0.5 mm thick will deflect about 0.01 mm when the pressure difference is 2.5 MPa (about 250 atm), and the resonance frequency will be about 500 Hz.

Equations (6.1), (6.2) and (6.3) are adequate for design purposes if there is no possibility of the centre of the diaphragm deflecting by more than $0.5t$. Larger deflections will produce nonlinearity, since a stretching action is added to the basic bending of the diaphragm, causing a stiffening effect. For large deflections (i.e. where $d_m > 0.5t$) eqn (6.4) should be used [1]

$$p = \frac{16Yt^4}{3R^4(1-v^2)} \left[\frac{d_m}{t} + 0.488 \left(\frac{d_m}{t}\right)^3 \right] \qquad (6.4)$$

(where R is the diaphragm radius $= D/2$). **N.B.**: Unlike eqns (6.1)–(6.3), where the units are those specified on Fig. 6.6, eqns (6.4), (6.5) and (6.6) require SI units.

A diaphragm such as that shown in Fig. 6.6, clamped at the edges, and subjected to a uniform differential pressure p, has at a radius r from the centre on the low-pressure surface a radial stress s_r and a circumferential stress s_t given by the equations:

$$s_r = \frac{3pR^2v}{8t^2}\left[\left(\frac{1}{v}+1\right)-\left(\frac{3}{v}+1\right)\left(\frac{r}{R}\right)^2\right]$$

$$s_t = \frac{3pR^2v}{8t^2}\left[\left(\frac{1}{v}+1\right)-\left(\frac{1}{v}+3\right)\left(\frac{r}{R}\right)^2\right].$$

(6.5)

The deflection d_r at any radius r is given by eqn (6.6):

$$d_r = \frac{3p(1-v^2)(R^2-r^2)^2}{16Yt^3}.$$

(6.6)

Equations (6.5) and (6.6) all give linear relations between stress and pressure, and are sufficiently accurate when $d_m \leqslant 0.5t$. Equation (6.6) can be used to estimate the degree of nonlinearity in any given application.

Figure 6.7 shows the form of the radial and tangential stress distributions on the diaphragm surface. Since regions of both positive and negative stress exist a bridge containing four active sensors may be used. This has two benefits. First, it provides first-order temperature compensation, which makes the pressure sensor output almost immune to thermal changes. Second, the resistance changes experienced by the individual strain gauges are additive in a four-arm bridge, which increases the sensor output as discussed earlier. Figure 6.8(a) shows how gauges 2 and 4 are placed as close to the centre of the diaphragm as possible, and are oriented to read tangential (tensile) strain. Gauges 1 and 3 are placed radially, close to the edge of the diaphragm, where the stresses are

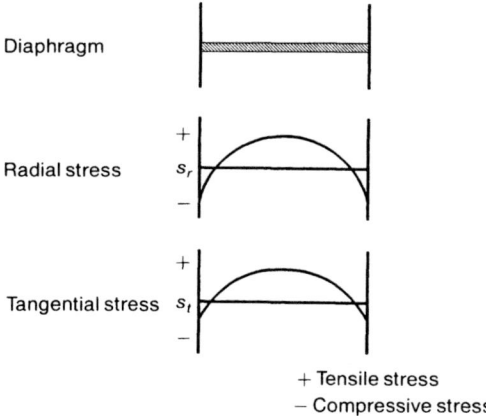

Fig. 6.7 Diaphragm stress distributions.

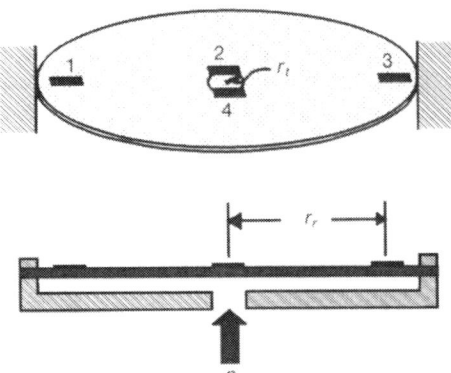

Fig. 6.8(a) Diaphragm and strain gauge pressure transducer.

compressive. Note that eqns (6.5) cannot be used directly to determine
the strains experienced by the gauges, since the diaphragm is in a state
of biaxial stress, and both the radial and tangential stress contribute to
the radial and tangential strain at any point. The general biaxial stress–
strain relation gives:

$$\varepsilon_r = \frac{s_r - vs_t}{Y}$$

$$\varepsilon_t = \frac{s_r - vs_r}{Y}.$$

Once the gauge strains are known the individual gauge resistance changes
ΔR can be obtained from the gauge factors.

To aid in the construction of miniature pressure transducers the discrete strain gauges shown in Fig. 6.8(a) can be replaced by a pressure-
diaphragm rosette such as that shown in Fig. 6.8(b). Rosettes of this form
are configured to take advantage of the radial strains at the diaphragm
edge and the tangential strains at the diaphragm centre. The solder connection points (I–VI) are placed in a low-strain region to avoid damaging
the joints.

6.2.4 Diaphragm pressure sensor fabrication techniques

Most small commercial pressure sensors are of the diaphragm type. Three
manufacturing techniques are used, two of which are based on stainless
steel diaphragms and one which uses silicon. Clearly stainless steel has
the advantage of ruggedness and is able to maintain a pressure seal even

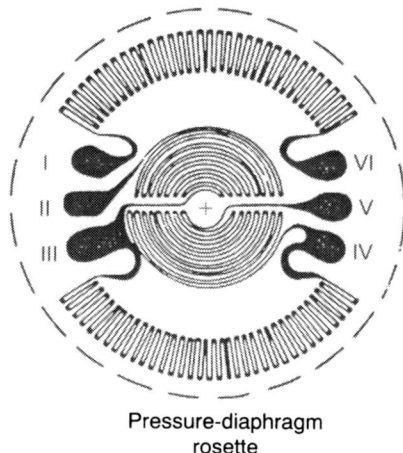

Pressure-diaphragm
rosette

Fig. 6.8(b) Strain gauge rosette for pressure transducers.

after electrical failure has occurred. For high-quality sensors where small size and low cost are not prime requirements, foil strain gauges are bonded to a stainless steel diaphragm. The diaphragm is usually encapsulated in a welded stainless steel enclosure, which may also contain the signal conditioning circuitry. In this case the assembly is usually referred to as a *pressure transducer* rather than a pressure sensor.

A cheaper approach is to use thick film piezoresistors on a stainless steel diaphragm [2]. This results in low-cost sensors which are physically smaller than the traditional bonded-foil strain gauge type, although not as small as silicon versions. It is difficult to apply thick film strain gauges to diaphragms less than about 3 mm in diameter. The thick film approach has a further advantage, in that the cost of setting up a thick film production plant is considerably less than that of a comparable silicon facility. If the required number of sensors runs to only hundreds or a few thousands per year, thick film will probably prove to be cheaper than silicon. The pressure sensor (i.e. the membrane and its associated piezoresistors) can be made to form part of a *thick film hybrid*, which allows signal conditioning circuits to be included within the sensor housing.

If very large numbers (> 100 000 say) of sensors are required, or if small size is essential, silicon architecture is almost always adopted. The predominance of silicon as the material used for creating low-cost sensors is partly due to the wealth of experience on manipulating and using silicon built up by the semiconductor industry. Its mechanical properties also make it well-suited for sensor use. Silicon has a density less than that of aluminium, and an elastic modulus approaching that of steel. It obeys

Hooke's law over a larger strain range than steel, and has a higher ultimate tensile strength. The temperature coefficient of expansion is of the same order as that of steel. Silicon's main disadvantage as a sensor material is that the complex technology used to create sensor architectures is expensive, and large production runs (typically > 100 000 units) are normally required to cover the tooling-up costs. The other disadvantages of silicon diaphragms are that they are easily damaged by water and other chemicals, and that they tend to shatter if struck by small gas or liquid-borne particles. Since silicon is a brittle crystalline material it will crack or shatter if exposed to shock loads. Protective screening techniques are available for shock protection, but these cannot always be used conveniently. Silicon pressure sensors are available using both piezoresistive and capacitance transducers.

6.3 Capacitance pressure sensors

The majority of silicon pressure sensors are piezoresistive and use diffused or ion-implanted piezoresistors. However, a number of successful capacitive pressure sensors have been used, for example by the Ford Motor Company with their silicon capacitance absolute pressure (SCAP) sensor used for engine control [3].

In piezoresistive pressure sensors diffused or implanted piezoresistors are arranged in single, half or full bridge forms. The differential strains sensed by the bridge resistors are used to achieve a push-pull effect to increase the sensitivity as discussed earlier. However, the limiting value of $\Delta R/R$ (i.e. change in resistance over unstrained resistance) is only about 10^{-2}, due to hysteresis and linearity considerations. PR devices are sensitive to transverse stressing of the sensor package, and the base strain sensitivity can be considerable. This increases the chance of interference, especially when the measurand is small and the sensor is subjected to vibration.

Capacitive pressure sensors are less prone to error. The diaphragm deformation caused by an applied pressure is turned directly into a change in capacitance, which can then be converted into an electrical form such as frequency, charge or voltage. Since the total change in capacitance is the integral of the changes produced by each part of the diaphragm, the susceptibility to side-stress is lower than that of a comparable PR pressure sensor.

Capacitance pressure sensor (CPS) designs can be grouped into three categories as shown in Fig. 6.9. Figure 6.9(a) shows a pedestal-and-ring arrangement, originally designed for a medical application [4]. The circular diaphragm is 610 μm in diameter and 10 μm thick.

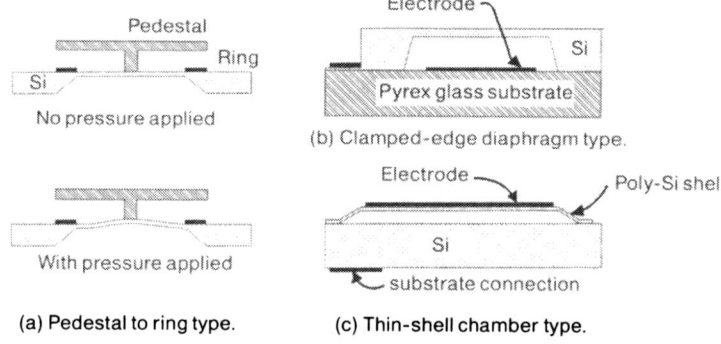

(a) Pedestal to ring type.

(b) Clamped-edge diaphragm type.

(c) Thin-shell chamber type.

Fig. 6.9 Capacitive pressure sensor structures.

The pedestal/ring gap is 5 μm and the zero pressure capacitance 1 pF. At 300 mmHg full scale $\Delta C/C = 0.2$ (c.f. 0.01 for PR strain gauges). The device is somewhat nonlinear (around 10% maximum), but is highly repeatable.

Figure 6.9(b) shows a different approach in which a silicon diaphragm is etched from a 300 μm thick wafer. The diaphragm is bonded to a Pyrex glass substrate. The gap chamber is either evacuated and sealed for absolute pressure measurement, or is supplied with a vent for gauge or differential pressure measurement. The Ford SCAP sensor discussed earlier is of this type. Some recent transducers have used silicon rather than glass as the substrate to reduce the thermal drift of the sensor and simplify the processing.

Figure 6.9(c) shows a thin shell pressure sensor. Micromachining techniques are used to deposit material on the silicon wafer and selectively etch off part of the material to construct a microchamber or cavity as shown in the diagram. The capacitor electrodes are on the top of the chamber and on the surface of the silicon substrate. With this approach no bonding is needed. However, care has to be taken to control built-in stress in the shell material.

In a diaphragm CPS the diaphragm deforms in response to the differential pressure p across its sides as shown in Fig. 6.10. This deformation changes the capacitance C between the electrodes, and a signal conditioning circuit is used to convert the output to a voltage V_o. If the supply voltage is V_s, the overall sensitivity of the CPS is S where:

$$S = \left(\frac{\Delta C/C_0}{p}\right)\left(\frac{\Delta V_o/V_s}{\Delta C/C_0}\right) = G_c A_V = \frac{\Delta V_o}{p V_s},$$

C_0 is the zero pressure capacitance. The first term:

$$G_c = \frac{\Delta C / C_0}{p}$$

is similar to the gauge factor of a piezoresistive sensor, and A_V is the capacitance-to-voltage gain of the signal conditioning circuit.

6.3.1 Capacitor microphones

The capacitor microphone (also called a condenser or an electrostatic microphone) is essentially a parallel-plate capacitor. One of the plates is a thin membrane exposed to the medium in which the sound is to be measured, so that pressure fluctuations alter the capacitor plate spacing. The resulting capacitance changes cause fluctuations in the voltage across the capacitor. The output signal consists of a varying voltage V_{out} as shown in Fig. 6.11.

The charge on the capacitor may be generated by an externally applied voltage, or by the properties of the material used to manufacture the

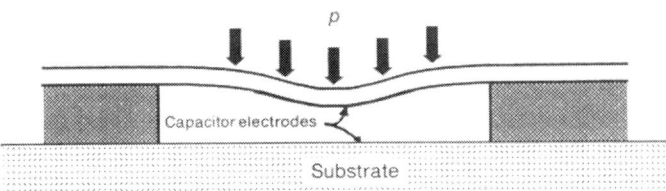

Fig. 6.10 Capacitive pressure sensor.

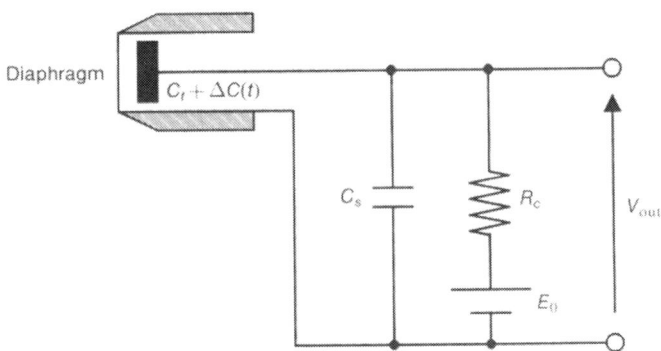

Fig. 6.11 Equivalent circuit of capacitor microphone.

capacitor. In the latter case the sensor is often called an electret micro-phone. An electret is a permanently charged insulating material made by allowing a molten plastic to solidify under the influence of a strong electric field.

The components of a typical capacitor microphone are shown in Fig. 6.12. A diaphragm made from metal foil is fixed close to an insulated rigid metal back plate. These two form a parallel-plate capacitor. A polarization voltage E_0 is applied across the plates as shown in Fig. 6.12. The polarization voltage source has a high impedance R_c, to ensure that the time constant $R_c(C_t + C_s)$ is long compared to the lowest sound frequency to be measured. A further reason for making the time constant long is that this ensures the charge stored on the capacitor is approximately constant. If the charge is constant and the capacitance varies, a varying voltage will appear across the plates. The capacitance C_s is due to unavoidable stray capacitance within the sensor, rather than a deliberate design feature.

If the sound pressure acting on the diaphragm produces a capacitance change ΔC, then the output of the microphone will be a voltage V_{out} where:

$$V_{out} = \frac{\Delta C \cdot E_0}{C_t + C_s},$$

since $C_t \gg \Delta C$. Note that the microphone sensitivity S is proportional to the polarization voltage E_0 but inversely proportional to the total

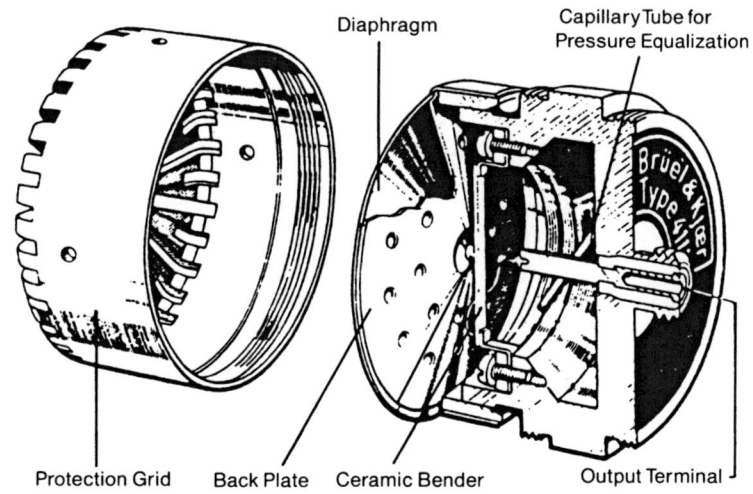

Fig. 6.12 Capacitor microphone construction (courtesy of B&K).

capacitance $C_t + C_s$. Capacitor microphones can be manufactured with sensitivities as high as 100 mV/Pa.

If we want the microphone to have a flat response then the capacitance change ΔC and hence the deflection of the diaphragm for a given sound pressure must be independent of frequency. In other words, the diaphragm must be 'stiffness controlled' [5], and it will have a natural frequency well above that of the highest frequency sound to be measured.

The voltage output from a capacitor microphone is proportional to the applied sound pressure. However, when the frequency of the sound increases to the point where the wavelength is of the order of the diaphragm diameter, the diaphragm acts as a high impedance obstacle in the sound field, from which the wave will be reflected. When this happens the pressure sensed by the diaphragm is incorrectly high. This is a manifestation of the 'pressure doubling' effect, discussed in reference [5]. In such a situation a microphone with a flat pressure response would give an incorrect reading.

6.4 Pressure switches

Pressure switches are frequently required in engineering. These devices have the function of indicating when a preset pressure threshold is passed. Examples include engine oil pressure sensors, hydraulic, and pneumatic system sensors. Most pressure switches are diaphragm-based and rely upon the diaphragm deflection being sufficient to close (or open) a contact. The threshold pressure is dictated by the diaphragm geometry, and is not usually adjustable by the user. Silicon micromachined pressure switches are available which can be used in applications where size or cost restricts the use of conventional pressure sensors. Pressure switches are generally cheaper than pressure sensors, since no strain gauges or interconnections are needed. A typical device is shown in Fig. 6.13. The diaphragm is typically 1 mm across and is fabricated from silicon using micromachining

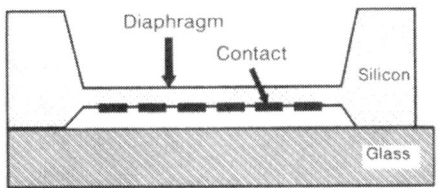

Fig. 6.13 Cross section of silicon pressure switch.

techniques. Instead of diffusing resistors into the silicon to measure strain, a metal pattern is deposited in an etched shallow well.

The setpoint (i.e. the pressure at which the contacts close) is adjusted during manufacture by a process known as electrical trimming. The metal pattern consists of a regular series of raised metal contacts as shown in the diagram. Clearly, when pressure is applied the contacts at the centre of the diaphragm will close first. As the pressure is increased contacts further away from the centre will close in succession. Each metal contact is manufactured to include a micromachined fuse, which can be blown by applying a voltage pulse. Thus if a switch is required to operate at low pressure, no fuses are blown. If a high-pressure switch is required some of the central fuses are blown, ensuring that a larger force has to be applied before a valid switch connection is made.

Pressure switches are generally more robust than pressure sensors, since the glass substrate provides an integral overforce protection for the silicon diaphragm.

The main disadvantage of this type of pressure switch is that the current through a pair of contacts must be smaller than that which is required to blow a fuse. For this reason commercial devices often include a transistor switching stage.

6.5 Pressure sensor environmental considerations

Pressure sensors often have to function in a particularly hostile environment. In automotive applications for example they may be subjected to extremes of temperature (typically from -40 to $+140\,°C$), vibration, electromagnetic interference, and chemical attack. Pressure measurements may be required in water, oil, hydraulic fluid, and air. Some of these fluids will attack some types of diaphragm, and special precautions may be necessary as a consequence. The sensor may be required to operate in a high-frequency, high-intensity electromagnetic environment. Electrical connections can also cause problems if small signals are to be transmitted. For these last two reasons it is desirable to include as many of the signal conditioning functions as possible within the sensor housing. The intention of this section is to indicate some of the likely areas in which problems may occur, and to outline some of the solutions.

6.5.1 Chemical attack

The diaphragm material which is most immune to chemical attack is undoubtedly stainless steel. Stainless steel diaphragms can readily be made with diameters as small as 3 mm and with a minimum thickness of 0.01 mm or less. Steel diaphragms may be electron-beam welded to the sensor housing to provide an impermeable and rugged seal. However, they are often strain-gauged manually, and this leads to a high unit cost which is often unacceptable.

Silicon diaphragms may be made cheaply by micromachining and can have piezoresistors diffused onto the surface as part of the manufacturing process. The unit cost of a silicon pressure sensor in mass production is very low, and they can be made extremely small with active diaphragm diameters down to 0.5 mm. The sensor package can include sophisticated signal conditioning functions, which are fabricated by the same processes used to form the diaphragm and its associated piezoresistors. However, silicon is attacked by water and many other fluids, and it is generally unwise to expose silicon directly to the fluid in which pressure is to be sensed.

Thick film piezoresistors on stainless steel offer a compromise between the approaches described above. The strain-gauging is to some extent automated, since a printing process is used, and the strain gauges are of lower cost than the bonded foil type. Signal conditioning circuits may be included by the creation of a *thick film hybrid device*. However, the size of a thick-film-on-steel pressure sensor is larger than that of a comparable silicon device.

A number of techniques are adopted which enable the user to profit from the benefits of silicon while also enjoying the ruggedness and immunity to chemical attack provided by stainless steel. The first of these involves the use of two diaphragms and silicone oil as a pressure transmitting medium. Figure 6.14(a) shows the arrangement. Silicone oil is chemically inert, and a silicon device can safely be exposed to it with no possibility

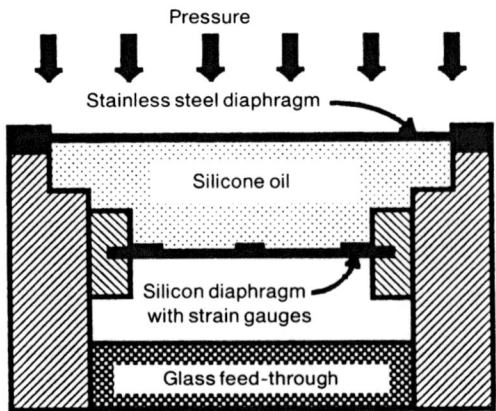

Fig. 6.14(a) Double diaphragm for chemical protection.

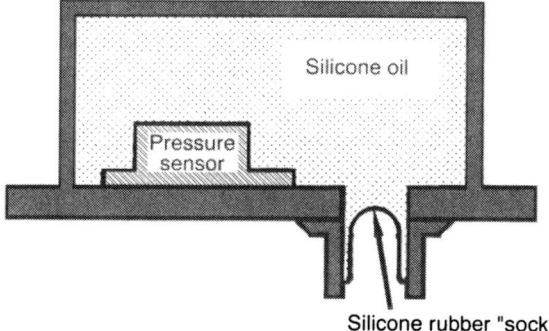

Fig. 6.14(b) Silicone rubber sock to protect sensor.

of damage. The silicon sensor is mounted within a stainless shell steel, sealed at the front by a welded stainless steel diaphragm. Pressure in the working fluid causes this diaphragm to deflect, which causes similar pressure variations to appear in the silicone oil. The silicon diaphragm senses the pressure in the silicone oil, rather than that in the (potentially damaging) working fluid. The electrical connections to the device are sealed by means of a glass feed-through.

An alternative approach is to use a flexible silicone rubber barrier (often called a 'sock') to contain the silicone oil as shown in Fig. 6.14(b).

6.5.2 Over range effects

Most pressure transducers will survive exposure to pressures beyond their measurement limit without damage. Manufacturers' data sheets usually give over range values, although these are generally very conservative and most devices will survive considerably more than the suggested limits. However, the effects of transducer resonance must be borne in mind when considering over range limits. If dynamic pressures are to be measured at frequencies more than about 30% of the resonance frequency, failure can occur as a result of the amplifying effect of mechanical resonance.

6.5.3 Pressure sensor acceleration sensitivity

If a diaphragm is clamped around its edges and the assembly accelerated, the diaphragm will deflect. The deflection is a function of the diaphragm mass and stiffness. In engineering applications any sensor acceleration which causes erroneous readings is likely to result from high-frequency vibrations. The acceleration levels resulting from, for example, vehicle motion are normally too low to affect the output of a pressure sensor.

Well-designed commercial pressure sensors have acceleration sensitivities of the order of 0.00005% of full-scale per g in the pressure sensitive direction. Cross-axis acceleration sensitivities are generally less than 20% of that in the sensitive direction.

6.5.4 Thermal sensitivity of pressure sensors

All pressure sensors have characteristics which are to some extent thermally-dependent. The sensitivity, linearity and zero offset can all change with temperature. The manufacturer of a pressure sensor will normally provide details of its thermal behaviour as part of the device's datasheet. Typical variations for silicon and stainless steel devices are of the order of 1–2% per 100 °C. This means that the sensitivity of the sensor will change by 1–2% away from the calibrated value for every 100 °C by which the sensor temperature departs from that at which it was calibrated. For example, a sensor with 1% thermal sensitivity and a nominal sensitivity of 1 mV/Pa will shift to between 0.99 and 1.01 mV/Pa if its temperature is changed by 100 °C.

Almost all sensors have some 'null pressure' or 'baseline' offset. This means that when the device is switched on the output will not be exactly zero when no pressure is applied. Zero offset values are usually specified in the datasheet, and are typically 10–20 mV for strain gauge bridge devices. Zero offsets can arise from incorrect balancing of the strain gauge bridge during manufacture, or from incorrect transducer mounting which results in the diaphragm being placed under stress. For example, over-torquing of screw-coupled pressure sensors can result in the appearance of an unexpected zero offset. It is common for large zero offsets to be present at the transducer output on switch-on, which reduce as the sensor 'warms up'.

Typical thermal changes in zero offset are of the order of 1–2% of full scale output per 100 °C. It should also be borne in mind that any thermal zero shift data supplied with a pressure sensor will only be valid for equilibrium temperature conditions. If large thermal gradients are present, or if rapid temperature changes occur, unpredictable output changes may take place.

The nonlinearity and hysteresis of a pressure sensor's output vary over the full-scale pressure range. Normally the manufacturer will provide 'worst-case' figures in the form of a percentage deviation from the ideal (straight line) output graph. It is rare for a manufacturer to specify the extent of any thermally-induced linearity change. However, these undoubtedly occur, and if extreme accuracy is required the user is well advised to carry out calibration checks at intervals across the required operating temperature range.

References

[1] *Instrumentation Reference Book*. by B.E. Noltingk. Butterworth, 1990.
[2] Thick-film sensors past, present and future. N.M. White and J.D. Turner. *Measurement Science & Technology*, Feb. 97, vol. 8, no 2.
[3] A miniature silicon capacitance absolute pressure sensor. by M.E. Behr, C.F. Bauer and J.M. Giachino. In *Proceedings of the International Automotive Electronics Conference*, I.Mech.E, 1981.
[4] Solid State Capacitive Pressure Transducers, by W.H. Ko. In *Sensors and Actuators*, vol. 10, (1986), pages 303–20.
[5] *Acoustics for Engineers*, by J.D. Turner and A.J. Pretlove. Macmillan, 1991.

7.1 Introduction and definitions

The subject of torque often seems to cause confusion, even amongst experienced engineers. Statements such as 'torque is the amount of twist in a shaft' or 'torque is a measure of the wind-up' are often heard. It is sensible therefore to begin with a formal definition of torque, before moving on to consider methods by which it can be measured.

Torque can be defined as a measure of the tendency of a force to rotate the body on which it acts about an axis. Everyday experience tells us that the 'rotating effectiveness' of a force increases with its perpendicular distance from the pivot. For example, when opening a door it is normal to push or pull as far as possible from the hinges, and we attempt to keep the direction of the pull or push perpendicular to the door.

The magnitude of the torque acting in a plane perpendicular to an axis is obtained by multiplying the force (or the component of the force in a plane perpendicular to the axis of rotation), by the perpendicular distance from the axis to the line of action of the force, as shown in Fig. 7.1. The SI units of torque are Newton-metres (Nm).

The description above applies to both static problems (where, for example, a structural member is subjected to a moment), and to the more important case (for engineers) of rotating shafts in torsion. For the engineer, torque in a rotating shaft is important, since torque produces strain, and strain (or at least the strain which the shaft will safely

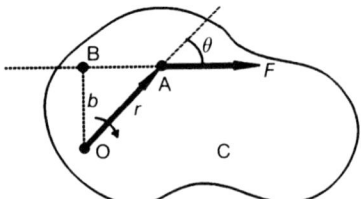

Force F acting on body C tends to rotate C about O.
Torque $\tau = Fb = Fr \sin \theta$
In vector notation torque $\tau = r \times F$.

Fig. 7.1 Definition of torque.

withstand) is the limiting factor in determining how much power the shaft can transmit.

The equations relating torque to power and strain are straightforward. A shaft rotating with angular velocity ω and carrying power P will undergo a torque T, where:

$$T = \frac{P}{\omega}. \tag{7.1}$$

A shaft of length l, polar moment of inertia J, and modulus of rigidity G, subjected to a torque T, will experience an angle of twist θ given by:

$$\theta = \frac{Tl}{JG}. \tag{7.2}$$

The maximum shear stress τ occurs at the surface of the shaft, and is given by:

$$\tau = \frac{Tr}{J}, \tag{7.3}$$

where r is the shaft radius as shown in Fig. 7.2. For a solid circular shaft the polar moment of inertia $J = \pi r^4 / 2$, so by substitution we have:

$$\theta = \frac{2Tl}{\pi r^4 G} \quad \text{and} \quad \tau = \frac{2T}{\pi r^3}. \tag{7.4}$$

Equations for the strains produced by torsion in shafts of other than circular cross-section are readily derived [1].

The most important application of torque measurement in many branches of engineering is in the assessment of mechanical power. Equation (7.1) shows how power can be readily be calculated from measurements of rotational speed and torque. Rotational speed measurement is relatively easy—often a white paint mark on the shaft or flywheel

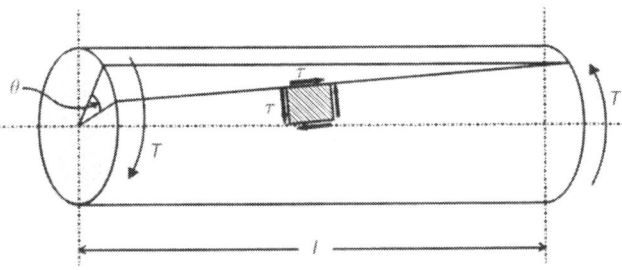

Fig. 7.2 Circular shaft (radius r) under tension. τ = shear stress.

and a simple optical reflectance sensor will suffice. Torque measurement is more difficult to arrange. Strain-gauged torque sensors (see section 7.2) are available, but are expensive, hard to use, and often unreliable in the long term.

7.2 Mechanical methods of torque measurement

One of the earliest (and still very useful) methods of measuring the torque in a rotating shaft uses a device known as an *absorption dynamometer*, in which all the power produced is absorbed by friction in a brake. This is the origin of the phrase *brake horsepower*, although shaft power is a less misleading term. A rope or belt brake is wrapped around a flywheel carried by the shaft, and is often water-cooled. The rope passes once around the flywheel and is attached to a mass M at the bottom as shown in Fig. 7.3. The other end of the rope is connected to a spring balance which measures the tension in the rope, T. The force in the lower end of the rope arises from the weight, and is Mg. If the spring balance reading is T, the difference in tension between the ends of the rope is $(Mg - T)$. If the radius of the flywheel is R, the torque will be:

$$\text{torque} = (Mg - T)R,$$

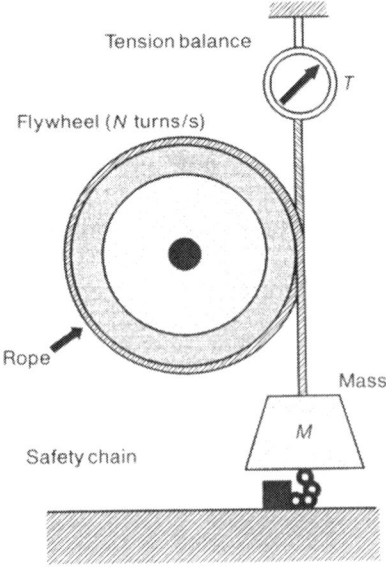

Fig. 7.3 Brake dynamometer.

and the power:

$$\text{power} = 2\pi N(Mg - T)R(\text{W})$$

(where N is the number of flywheel revolutions per second).

The danger inherent in this arrangement is that the brake may jam, throwing the weight over the top of the flywheel. To avoid this alarming possibility a strong safety rope or chain is always used as shown in Fig. 7.3, which prevents the weight being lifted more than a few centimetres.

A more sophisticated brake dynamometer frequently used for torque measurement is the hydraulic type, originally invented by W. Froude. In a Froude dynamometer (see Fig. 7.4) the energy from a rotating shaft is transferred to kinetic energy in water, which is then brought to rest.

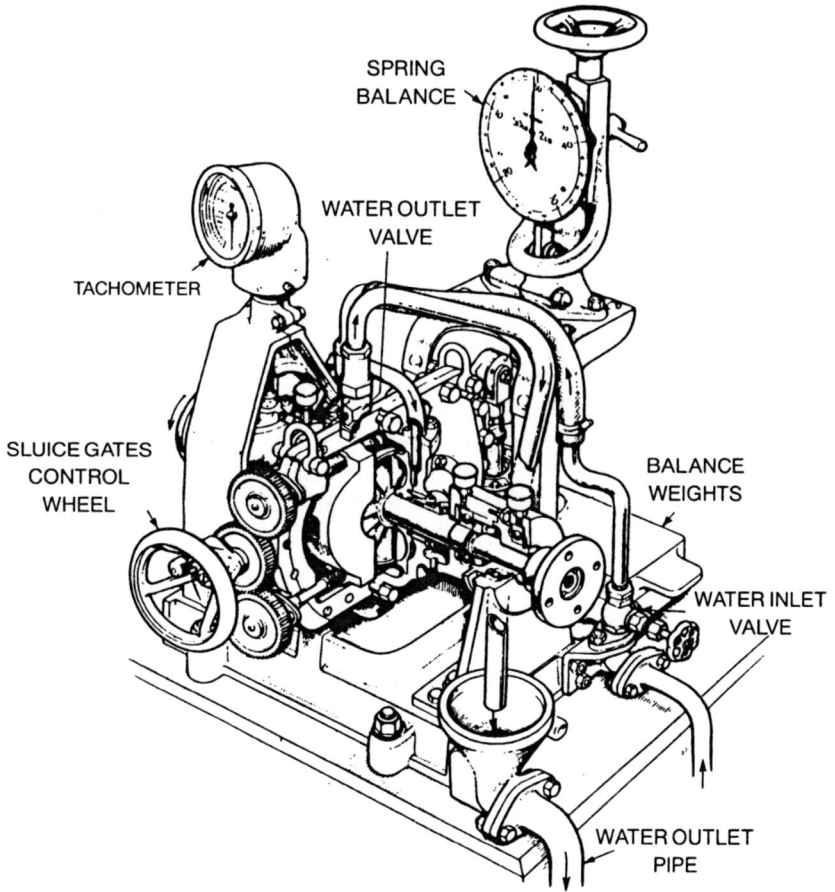

SPRING
BALANCE

WATER OUTLET
VALVE

TACHOMETER

SLUICE GATES
CONTROL
WHEEL

BALANCE
WEIGHTS

WATER INLET
VALVE

WATER OUTLET
PIPE

Fig. 7.4 Froude water-brake dynamometer.

The torque required to restrain the device is measured by (usually) a spring balance. The advantage of a Froude dynamometer is that unlike a rope brake device there is no possibility of it 'snatching'. However, Froude dynamometers are more expensive than rope brakes.

It is interesting to note in passing that the automotive fluid flywheel was almost certainly developed from the Froude dynamometer.

An entirely mechanical method of torque measurement is based on a measurement of the force required to restrain a gearbox. As far as the authors are aware it has not been widely applied, although there seems to be no reason why it should not be successfully used. Any gearbox which changes the rotational speed of a shaft will change the torque in inverse proportion (assuming friction can be neglected). The *ratio* of the input torque T_{in}, to output torque T_{out} is equal to the reciprocal of the speed ratio, and the *difference* between the input and output torques is the torque needed to restrain the gearbox. Thus:

$$\frac{T_{in}}{T_{out}} = \frac{\omega_{out}}{\omega_{in}} \quad \text{and} \quad (T_{in} - T_{out}) = T_{restraining}.$$

By measuring the input and output speeds, and the restraining torque T_{in} and T_{out} can be calculated. This approach has been found to be very useful in university laboratories, where costs are of overriding importance, since it allows a torque measuring system to be improvised cheaply using a scrap back axle from a rear wheel drive vehicle as shown in Fig. 7.5. The engine or other mechanical power source is connected to one wheel shaft, and the load to the other. The propeller shaft coupling is locked to the differential housing. The torque required to restrain the housing is measured with a spring balance or electronic force transducer, and is twice the torque being transmitted through the system, since the input and output shafts revolve in opposite directions.

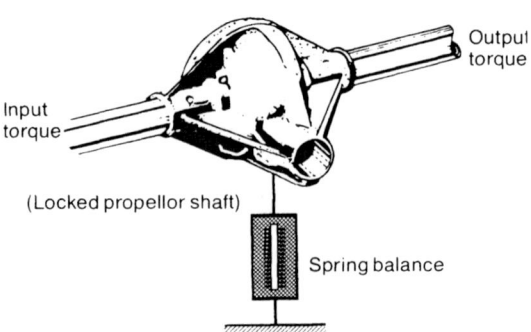

Fig. 7.5 'Back-axle' torque transducer.

7.3 Strain gauge torque transducers

Torque transducers based on strain measurement are normally made by applying strain gauges to a shaft to measure the shear strain caused by torsion, as shown in Fig. 7.2 and discussed in section 7.1. This type of transducer is widely used and probably forms the most common type of torque sensor. The major disadvantage of this approach is that additional equipment is usually required to transmit power to the rotating shaft and energize the strain gauge bridge, and also to retrieve the data. This apparatus can take the form of a set of slip rings, rotary transformers, or battery-powered radio telemetry equipment. Regardless of which is chosen, the need for some form of power and/or data transmission system makes the measurement of torque more expensive than, say, that of pressure or temperature. In addition, slip rings (and to some extent rotary transformers) can be unreliable when operated in a dirty environment, and may be prone to radio frequency interference (RFI).

The shear stresses illustrated in Fig. 7.2 cause strains to appear at 45° to the longitudinal axis of the shaft. The conventional arrangement of strain gauges for torque measurement is shown in Fig. 7.6. The gauges must be placed precisely at 45° to the shaft axis, otherwise the arrangement is sensitive to bending and axial stresses in addition to those caused by torsion. Accurate gauge placement is facilitated by the availability of special 'rosettes', in which two gauges are precisely positioned on a common backing as shown in Fig. 7.7. The use of four active strain gauges in a bridge arrangement gives complete thermal compensation [3].

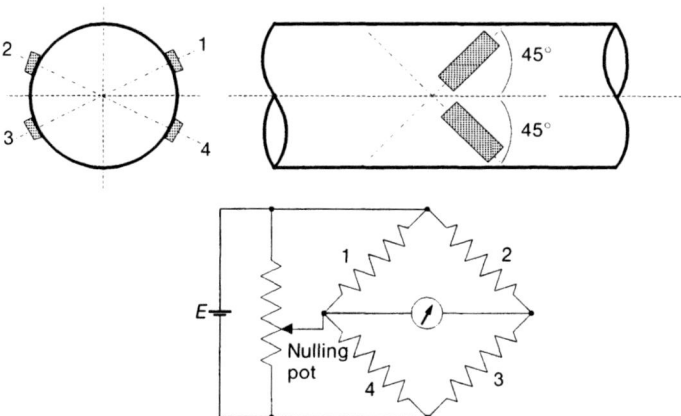

Fig. 7.6 Strain gauges for shaft torque measurement.

Figure 7.6 shows an arrangement of strain gauges on a solid circular shaft. The same gauge positioning can be used on a hollow circular shaft. When torque is applied to a thin-walled cylinder the shear stress is assumed to be constant throughout the wall [4]. In such cases it is often convenient to place the strain gauges on the inner surface of the tube, where they are afforded a degree of mechanical protection.

Shafts of other than circular cross-section are sometimes used for torque measurement as shown in Fig. 7.8. For measuring low levels of torque the cruciform or hollow cruciform configuration is sometimes used.

Fig. 7.7 Strain gauge rosette for torsion.

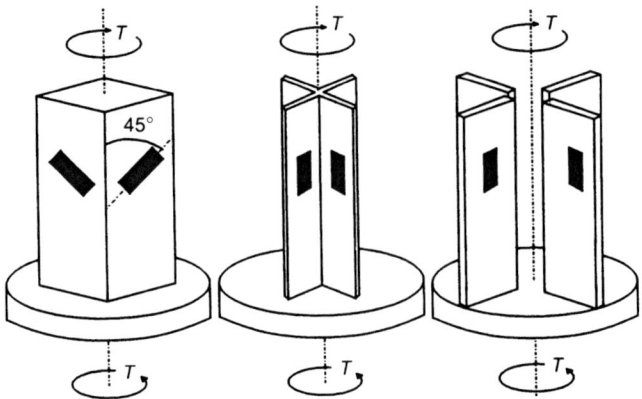

Fig. 7.8 Torque sensor designs.

A solid square shaft is suitable for larger torque values, and has a number of advantages over the circular shaft of Fig. 7.6. The strain gauges are more easily aligned and attached to a flat surface, and since the corners of a square section in torsion are stress-free [5], they provide a good location for the solder joints between leads and strain gauges. These joints are often a source of unreliability due to fatigue failure if they are located in a high-stress region. Finally, a square shaft is much stiffer in bending than a circular one of equivalent torsional stiffness, so the effects of bending (which will appear if the gauges are misaligned) are reduced.

7.4 Torsion bars

Torque in a shaft leads to elastic deflection. The resulting strain can be measured at a point as described in the preceding section, or alternatively the gross relative motion between the ends of the shaft may be used to indicate the torque. Just as in the case of strain gauge systems, a major difficulty is the necessity of being able to measure the deflection while the shaft is rotating. However, there are advantages in using shaft deflection. First, the need for precise location and orientation of the strain sensors is avoided. Second, since the effect of an applied torque is integrated along the length of the shaft, the influence of any local variation in material properties or shaft geometry is reduced. Third, the (relatively) larger displacements available when movements of the two ends of a shaft are compared make it possible to design a variety of non-contact torque measurement systems which avoid the need for slip rings.

Figure 7.9 shows a typical torsion-bar torquemeter using an optical method for deflection measurement. The relative angular displacement between the ends of the torque-transmitting member is read from the position of the pointer on disk 2 relative to the calibrated scale fixed to disk 1. The 'persistence' of human vision and the stroboscopic effect of intermittent viewing make it possible to operate this system from about 600 rpm (10 Hz) upwards.

A torsion-bar system using capacitive torque sensing has been demonstrated for automotive use [6]. An automotive drive shaft was fitted with a concentric sleeve of dielectric material as shown in Fig. 7.10. The sleeve was fixed to the shaft at one end, and rests on a rubbing bearing at the other end. When torque is applied to the shaft it causes relative motion between the surface of the shaft and the free end of the concentric tube. This motion is used to vary the capacitance between two opposing patterns of conducting strips, one of which was applied to the shaft and one to the tube.

The capacitive torque sensor was connected to an inductor coil wound around the shaft. The resulting passive circuit thus has a resonance frequency which depends on the applied torque. The passive resonant circuit rotates with the drive shaft, and is excited from an adjacent stationary location by inductive coupling using a second inductor coil driven by an oscillator as shown in Fig. 7.11. The problem of torque measurement then

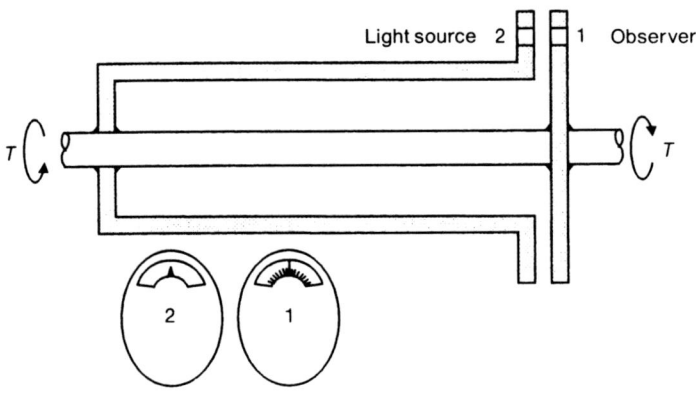

Fig. 7.9 Torsion-bar torque transducer.

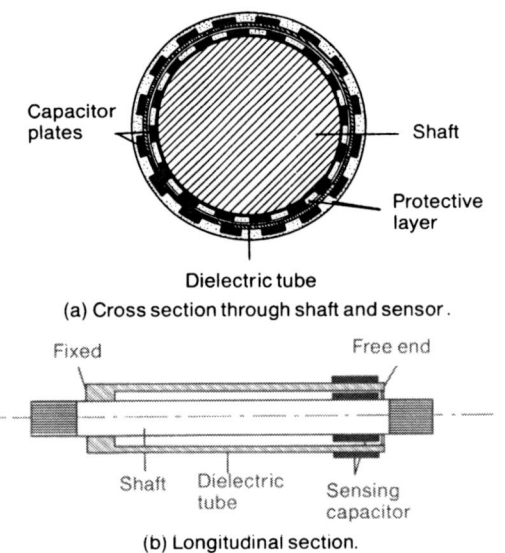

(a) Cross section through shaft and sensor.

(b) Longitudinal section.

Fig. 7.10 Construction of torque sensor.

becomes one of measuring the frequency at which resonance occurs. When the oscillator frequency is the same as that at which resonance occurs in the passive circuit an increased current is drawn. If the frequency at which this occurs is measured it can be used to indicate the torque. The advantage of this arrangement is that no physical connection between the rotating shaft and the vehicle body is required.

An optical torsion-bar sensor intended for use as part of an electric power assisted steering (EPAS) system has also been proposed by the Lucas Advanced Engineering Centre in Birmingham [7]. The sensor uses a pair of slotted disks positioned at the ends of a torsion bar as shown in Fig. 7.12. Light from a light-emitting diode (LED) passes through the

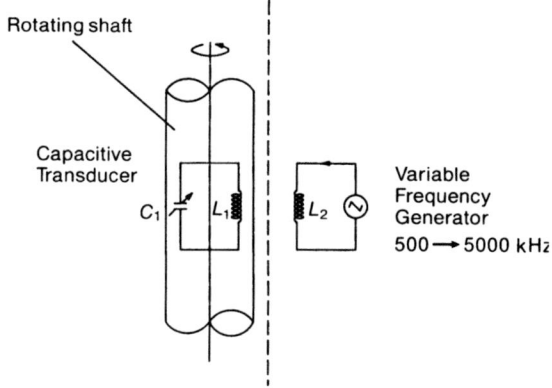

Fig. 7.11 Rotating resonant circuit excited by inductive coupling.

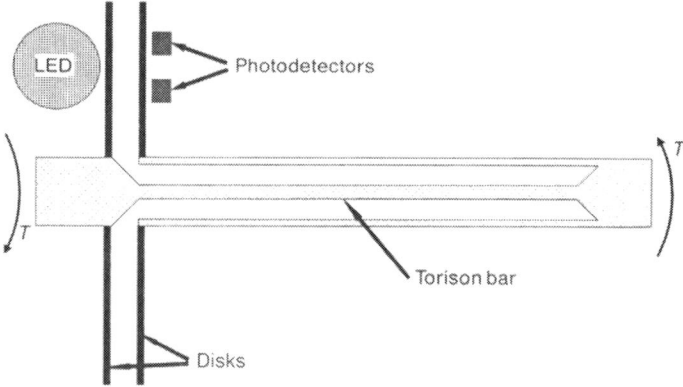

Fig. 7.12 Lucas EPAS ratiometric torque sensor.

slots in the disks and is received by a photodetector. Torque variations cause the amount of overlap between the disks to vary, and hence the output from the photodetector. However, the Lucas system exhibits a number of refinements which are intended to make it more suitable for automotive use. The most important of these is the use of a ratiometric technique to cancel out the effect of any variation in the source illumination intensity. The slotted disks are illuminated by a common LED as shown in Fig. 7.12. The amount of light emitted by the diode will vary if the supply voltage changes. Even if a well-regulated supply is available, the light output from an LED reduces by up to 40% as the device ages. The ratiometric effect is achieved by arranging for each slotted disk to carry two tracks of slots, positioned so that as torque is applied in one direction the light intensity transmitted through the outer track (A) increases, while that passing through the inner track (B) decreases. The light passing through each track is measured by a pair of photodiodes as shown in Fig. 7.12. The torque is calculated by measuring the outputs from photodiodes A and B and then evaluating the expression:

$$\text{torque} = \frac{A - B}{A + B}.$$

The magnitude of the result gives the torque, and the sign the direction (i.e. clockwise or anticlockwise) in which it is applied. Provided both channels are affected equally, this technique ensures that the torque measurement is independent of the source intensity. Furthermore, for a given source intensity the quantity $(A + B)$ should be a constant which is independent of torque, and this value can be used to check that the sensor is operating correctly. If $(A + B)$ moves outside preset limits an appropriate warning may be given. A selftest facility of this kind is obviously essential in a safety-critical system such as vehicle steering.

The main problem with the Lucas system appears to be that the geometry of the photodetectors, and their location with respect to the slots on the disks, is critical if ripple in the sensor output is to be prevented as the disks rotate. Variations in the output can only be avoided if the sensitive area of the light detectors corresponds exactly to an even multiple of the slot area. Reference [7] proposes the use of masks to give the correct detection area and to collimate the light source. The ripple amplitude after these improvements is reported to be better than 10% of the full-scale measurement range. Although this level of accuracy would not be acceptable for a laboratory torque sensor, it is probably adequate for power steering applications.

Work on measuring the twist or 'wind-up' along the crankshaft of an engine using slotted disks at each end has also been reported [8]. However, the very high levels of torque variation which result from multicylinder operation are alleged to make it difficult to obtain accurate results.

7.5 Non-contact magnetic methods

A number of torque sensors utilizing the magnetostrictive effect have been reported. A good example of this approach is a device described in reference [8] and shown in Fig. 7.13. Magnetostriction is an effect which occurs in ferromagnetic materials such as steel, where the magnetic permeability is affected by stress. Equation (7.3) shows that the stress in a shaft is proportional to the applied torque, and it follows that torque must change the permeability of the shaft if it is made of a magnetic material such as steel. The effect is small but can be measured by an arrangement such as that shown in Fig. 7.13. The torque sensor consists of five coils arranged as shown, wound onto a common five-armed core. The centre coil can be thought of as the primary winding of a transformer, and the four circumferentially-positioned coils act as secondaries. Magnetic coupling between the primary and the secondaries is provided by the steel shaft, which is positioned close to the sensor as shown. The primary coil is excited by an AC current, and produces an oscillating magnetic field within the shaft. The four secondary coils are connected together in a Wheatstone bridge arrangement, and are positioned so that they lie over the lines of principal stress, which follow a helical path at 45° for a cylinder in torsion. When the shaft is not under torsion equal currents are induced in the four secondaries and the bridge out-of-balance

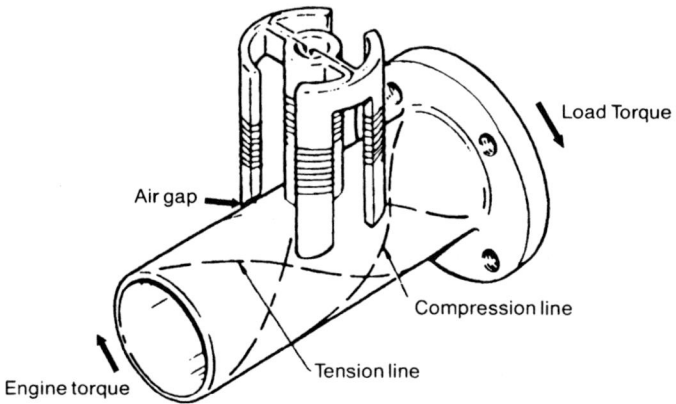

Fig. 7.13 Magnetostrictive torque sensor. (With acknowledgements to Spectrol Electronics.)

voltage is zero. When torque is applied to the shaft the permeability in the tension and compression directions will change by equal but opposite amounts, and the amplitude of the resulting bridge output voltage is proportional to the applied torque.

There are four main problems with this type of torque sensor. They are:

(1) inhomogeneity of the shaft material,
(2) sensitivity to changes in the sensor/shaft gap,
(3) thermal effects, and
(4) variations in the sensor output due to changes in the shaft rotation speed.

The first of these effects is the most serious. The permeability of the material from which the shaft is made can vary by up to 50% around the circumference of the shaft. For a constant torque the output signal from the sensor can 'ripple' as a result at a frequency equivalent to the rotation rate. This characteristic makes it very difficult to measure instantaneous torque levels around a rotating shaft. However, the use of smoothing circuits allows the device to be used for measuring the average torque in the shaft by integrating over several revolutions.

7.6 Surface acoustic wave (SAW) torque transducers

SAW devices are based on a theory propounded by Lord Rayleigh in 1885 [10], which showed that waves (known as surface waves or Rayleigh waves) can propagate along the surface of an isotropic elastic medium. Surface acoustic waves can be excited and detected using piezoelectric transducers etched with a pattern of interdigitated electrodes as shown in Fig. 7.14. The frequency at which the SAW device operates is determined

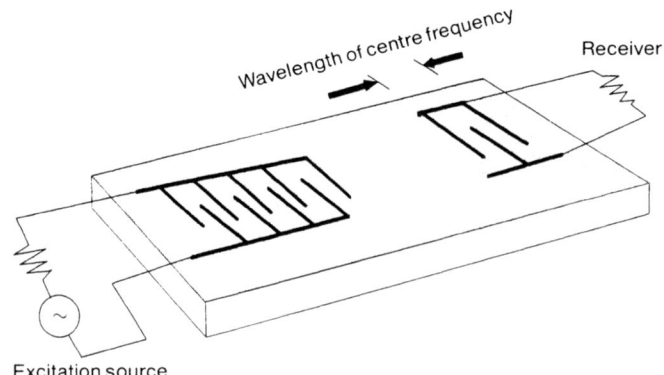

Fig. 7.14 SAW electrode arrangement.

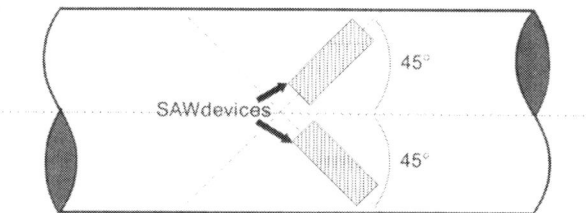

Fig. 7.15 SAW transducers for shaft torque measurement.

by the electrode geometry. For torque measuring, two SAW devices are attached to the shaft undergoing torsion. The shear strain resulting from torque (see Fig. 7.2) changes the geometry of the electrodes, and hence the operating frequency of the device. At a frequency of 500 MHz 1000 με (microstrains) will alter the SAW frequency by 500 kHz [11]. The two transducers are positioned on the shaft at 45° as shown in Fig. 7.15. Each transducer forms part of the feedback loop in an oscillator, such that the output frequency is a function of the SAW geometry. The two SAW transducers are used in a half bridge configuration, one undergoing tension and the other compression. The resulting two frequencies are added or subtracted: the difference in frequency gives a measure of torque, and the sum can be used to estimate temperature [11].

The SAW devices can be driven without the need for any electrical connection if capacitive or inductive pickups are used [11]. This feature makes SAW-based systems particularly attractive for torque measurement, since (as discussed in section 7.3) telemetry based on slip rings or rotary transformers can be a source of unreliability.

References

[1] *Formulas for stress and strain*. R.J. Roark and W.C. Young. McGraw-Hill.
[2] Automotive Transducers: An Overview. M.H. Westbrook. In *Proceedings of the IEE* Part D, volume 135 no. 5, pages 339–47, September 1988.
[3] *Instrumentation for Engineers*. J.D. Turner. Macmillan, 1988.
[4] *Mechanics of Engineering Materials*. P.P. Benham and R.J. Crawford. Longman, 1987.
[5] *Theory of Elasticity*. S.P. Timoshenko and J.N. Goodier. 3rd edition, McGraw-Hill, 1982.
[6] The development of a thick-film non-contact shaft torque sensor for automotive applications. J.D. Turner. *J. Phys. E: Scientific instrumentation*, volume 22 (1989), pages 82–8.

[7] Application of an optical torque sensor to a vehicle power steering system. R.J. Hazelden. In *Proceedings of the IEE* colloquium C12: 1992/107, May 1992.

[8] Sensors for automotive applications. M.H. Westbrook. *J. Phys. E: Sci. Instrumentation*, volume 18, 1985, pages 751–8.

[9] *Non-contacting sensors for automotive applications*. R.F. Wells. SAE Technical Paper Series no. 880407, (1988). SAE, 400 Commonwealth Drive, Warrendale, PA, USA.

[10] On waves propagated along the plane surface of an elastic solid. Lord Rayleigh. In *Proceedings of the London Mathematical Society*, volume 7, pages 4–11, 1885.

[11] *Acoustics sense torque at low cost*. T. Shelley. Eureka, September 1993, pages 48–9.

Flow sensors

8.1 Introduction

Flow sensors are important in many engineering applications, particularly those where the measurements are used to control a manufacturing process. Flow sensors may also be found in many consumer products, such as cars, where measurements of instantaneous fuel flow are used by the engine management computer. The subject of flow measurement can be sub-divided into vector flow, where the direction and the magnitude of flow are sensed, volume flow, and mass flow rate. Mass flow rate transducers will not be described separately, since they generally consist of a volume flow sensor combined with a density measurement (in gases, density is obtained from a measurement of temperature [1]).

There are a very large number of flow sensors available which space precludes us from considering. Reference [2] gives further details of some of the more advanced techniques. The following descriptions cover the most commonly-used laboratory devices:

8.2 Vector flow transducers

The two most common methods of measuring vector flow are by means of hot-wire anemometers or by the use of pitot tubes.

8.2.1 Hot-wire and hot-film flow transducers

A hot-wire anemometer consists of an electrically heated wire or metallized-film probe, which is placed in the moving fluid as shown in Fig. 8.1. The rate at which thermal energy is lost from the wire depends on the flowrate as shown in eqn (8.1):

$$I^2 R = hA(T_w - T_f),$$
(8.1)

where: I is current,

R the electrical resistance of the wire,

h is the wire or film coefficient of heat transfer,

A the probe's heat transfer area,

T_w the wire (or film) temperature,
T_f the temperature of the moving fluid.

The film coefficient of heat transfer h is a function of the fluid velocity v and two constants, C_0 and C_1:

$$h = C_0 + C_1\sqrt{v}. \tag{8.2}$$

The output of a hot-wire anemometer is also a function of the direction of flow. A simple transducer such as the device shown in Fig. 8.1 has a maximum output when the flow is orthogonal to the wire, as shown in Fig. 8.2, and the output reduces in an approximately cosine fashion as

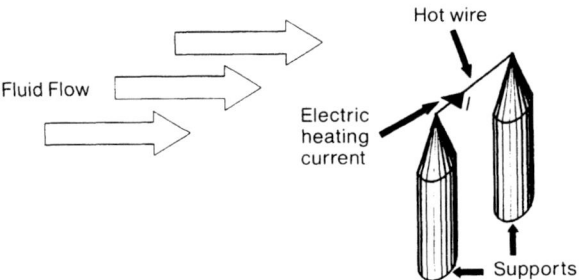

Fig. 8.1 Hot wire anemometer probe.

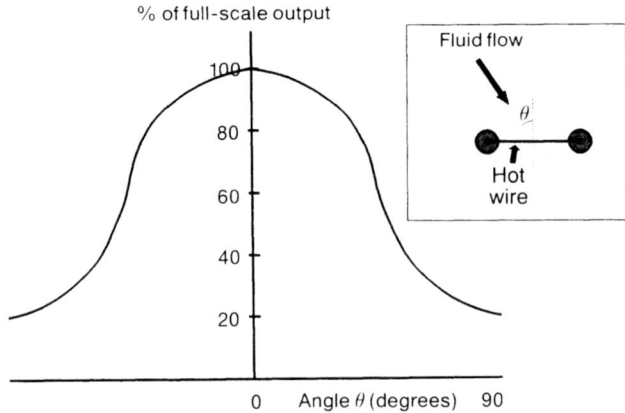

Fig. 8.2 Approximate directional response of hot-wire anemometer.

the probe is rotated, reaching a minimum of around 20% of the maximum output when the wire is aligned with the flow as shown in Fig. 8.2.

The wire has a known resistance/temperature characteristic, and is made to form one arm of a bridge circuit (see chapter 5). Either the heating current is maintained at a constant value, or a feedback system is used to keep the temperature of the wire constant. This second approach has the advantage that the heating current is then proportional to the flow.

8.2.2 Pitot tube flow sensors

Figure 8.3 shows the arrangement used to measure vector flow with a pitot tube. The device actually consists of two concentric tubes, the inner one having its open end facing the oncoming flow. The outer tube is closed at its end but has a number of holes in the walls. Both tubes contain the same fluid as the flowing medium. For the measurement to be success-ful, the direction of the flow must be known and the pitot tube aligned with this direction as shown in Fig. 8.3. The pressure in the outer tube p_{stat} is the static pressure in the fluid. The pressure in the inner tube (p_{stag}) is the sum of the static pressure and a pressure due to the impact of the flow on the stationary fluid in the tube. Thus, the flow generates a pres-sure differential across the two parts of the pitot tube, which is shown being measured by a manometer in Fig. 8.3. Obviously, the magnitude of the pressure differential depends on both the flow rate and direction. Assuming the flow is steady and one-dimensional, and that the fluid is

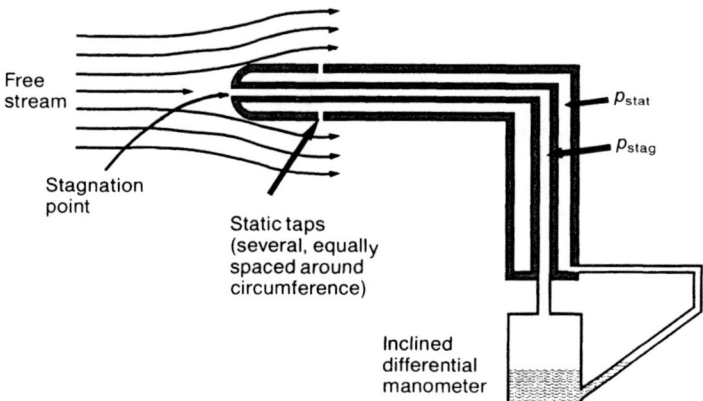

Fig. 8.3 Pitot tube flow transducer.

incompressible and frictionless [3]:

$$V = \sqrt{\frac{2(p_{stag} - p_{stat})}{\rho}} \qquad (8.3)$$

8.3 Volume flow sensors

8.3.1 Orifice plates

Most flow sensors act as modifiers. A common form comprises perforated plates or nozzles which partly restrict the flow to produce a pressure drop, which can then be measured. The process is analogous to estimating the flow of electric current by measuring the voltage drop across a small resistor, through which the current is made to flow. The sharp-edged orifice shown in Fig. 8.4(a) is probably the most common type, since it is simple and cheap to construct. If one-dimensional flow of an

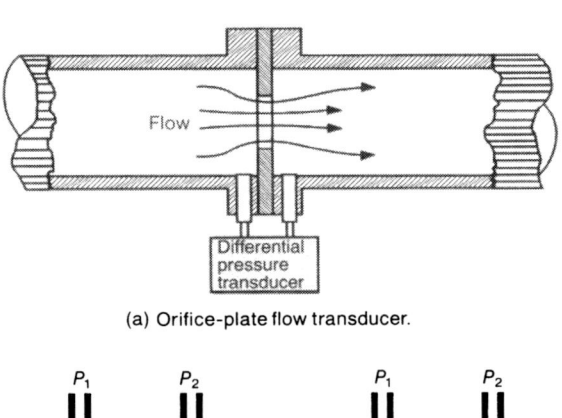

(a) Orifice-plate flow transducer.

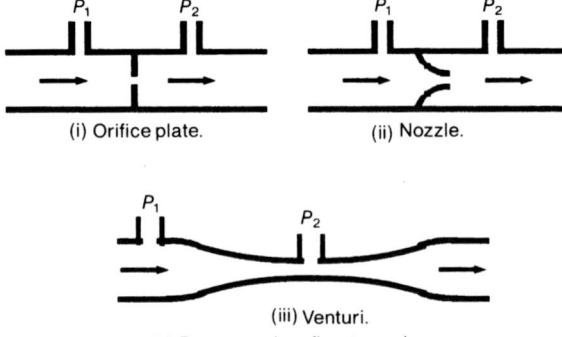

(i) Orifice plate.　　　　　(ii) Nozzle.

(iii) Venturi.

(b) Pressure-drop flow transducers.

Fig. 8.4 Pressure drop flow sensors.

incompressible frictionless fluid is assumed, theory gives the following result [3]:

$$\text{flowrate } Q_t = \frac{A_{2f}}{\sqrt{1-(A_{2f}/A_{1f})^2}} \cdot \sqrt{\frac{2(p_1-p_2)}{\rho}}, \tag{8.4}$$

where A_{1f}, A_{2f} are the cross-section flow areas (in m^3) where p_1, p_2 are measured,

ρ is the fluid density (kg/m^3),

p_1, p_2 are static pressures (Pa).

8.3.2 Turbine flowmeters

The drawback of all modifier flow sensors is that they disturb the flow by their presence. Turbine flow sensors are more satisfactory, in that the amount of disturbance they cause is smaller. A turbine flow sensor consists of a small vaned device which is placed in the flow. Figure 8.5 shows a typical example. The moving fluid causes the turbine to rotate, and the rotation rate is usually sensed externally to the flow (by, for example, a Hall effect sensor).

8.3.3 Rotameters

A rotameter consists of a vertical tube with a tapered bore as shown in Fig. 8.6, in which a 'float' is placed. The float assumes a position which

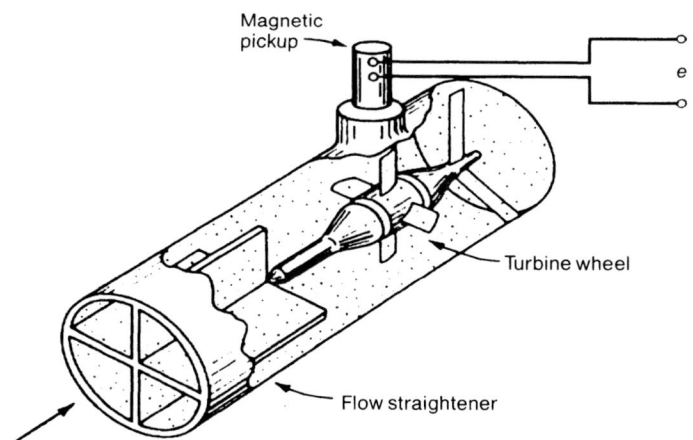

Fig. 8.5 Turbine flow sensor.

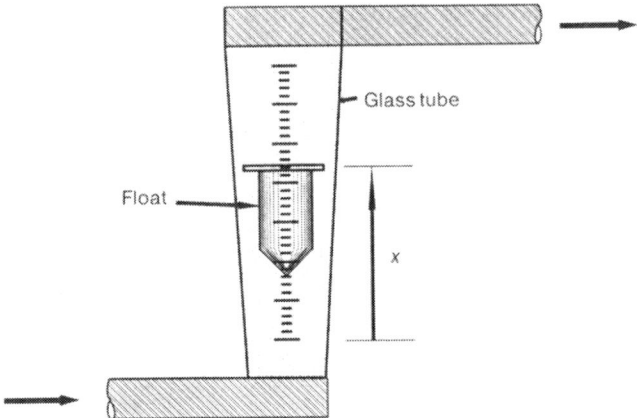

Fig. 8.6 Rotameter.

depends on the flowrate: if there is no flow, it sinks to the bottom of the tube. Flow causes an upward force on the float, and for a given flowrate the vertical forces of pressure, gravity, viscosity, and buoyance are balanced, giving a stationary float position. The equilibrium position is stable, since the meter flow area varies continuously with displacement; thus the device may be thought of as an orifice with variable area. The downward force (gravity minus buoyancy) is constant, so for stable equilibrium the upward force (pressure drop multiplied by float cross-sectional area) must also be constant. Since the float area is fixed, the pressure drop must be constant. For a fixed flow area the pressure drop Δp varies with the square of flowrate, so to keep Δp constant for different flow rates the orifice area (and hence float position) must vary.

8.4 Laser Doppler and correlation flow transducers

Figure 8.7 shows a simplified view of a laser Doppler flow transducer, which provides a non-invasive flow measurement technique for transparent fluids. A powerful main beam and a weaker reference beam are passed through the fluid at different angles, and some of the main beam is reflected in the direction of the reference beam by particles or small eddies within the fluid. These act as Doppler reflectors, causing a small frequency shift proportional to the component of velocity in the direction of the main beam. Because optical frequencies are very high, a beat frequency can be measured electronically, and flows over a range from a few millimetres to several hundred metres per second sensed.

Another laser-based flow measurement technique uses the principle of correlation. A beam of light is reflected from two adjacent points in the flow, as shown in Fig. 8.8, and the correlation function of the two signals $A_1(t)$ and $A_2(t)$ is determined. Provided the two reflection points are reasonably close, $A_2(t)$ will simply be a delayed version of $A_1(t)$, and the correlation function will have a peak corresponding to the transit time between the reflection points. Figure 8.8 shows the method applied to fluid in an open channel, where the surface ripples produce the signals $A_1(t)$ and $A_2(t)$. In closed pipes (which have to be transparent, or at least be provided with a transparent window) it is often necessary to introduce small particles of a material such as plastic powder which can be carried along by the flow to generate $A_1(t)$ and $A_2(t)$.

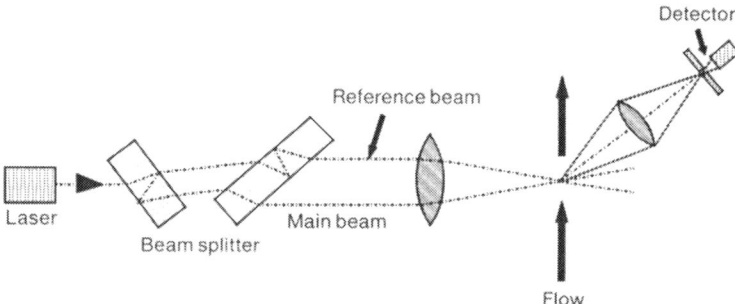

Fig. 8.7 Laser Doppler flow transducer (simplified).

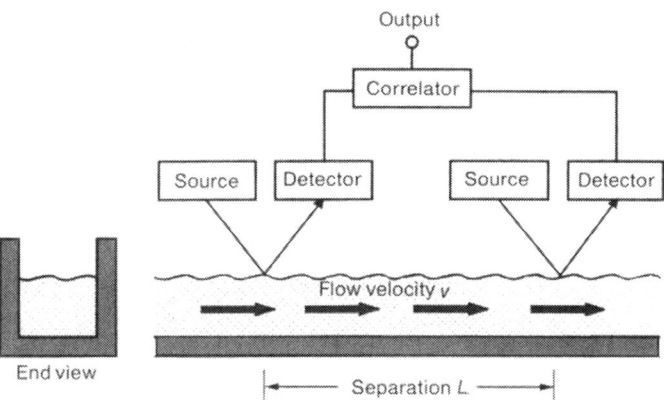

Fig. 8.8 Laser correlation flow measurement system.

8.5 Ultrasonic flowmeters

In an ultrasonic flowmeter small-magnitude pressure disturbances at a high frequency (normally 20–50 kHz) are propagated through the fluid at the speed of sound. Figure 8.9 shows a typical arrangement. If the fluid is flowing with velocity v_f, and the speed of sound in the stationary fluid is c, the propagation speed V will be the sum of the two, i.e.

$$V = v_f + c.$$

Figure 8.9(a) shows the most direct implementation of this approach. When the flow velocity is zero, the transit time t_0 is given by:

$$t_0 = \frac{L}{c},$$

where L is the distance between transmitter and receiver as shown in Fig. 8.9. In water the speed of sound $c \approx 1500$ m/s, so if $L = 10$ cm, $t_0 \approx 70\,\mu s$. If the fluid is moving at a velocity v_f, the transit time t becomes:

$$t = \frac{L}{c + v_f} = L\left(\frac{1}{c} - \frac{v_f}{c^2} + \frac{v_f^2}{c^3} - \cdots\right) \approx \frac{L}{c}\left(1 - \frac{v_f}{c}\right).$$

If the change in transit time $\Delta t = t_0 - t$, then:

$$\Delta t \approx \frac{Lv_f}{c^2}. \tag{8.5}$$

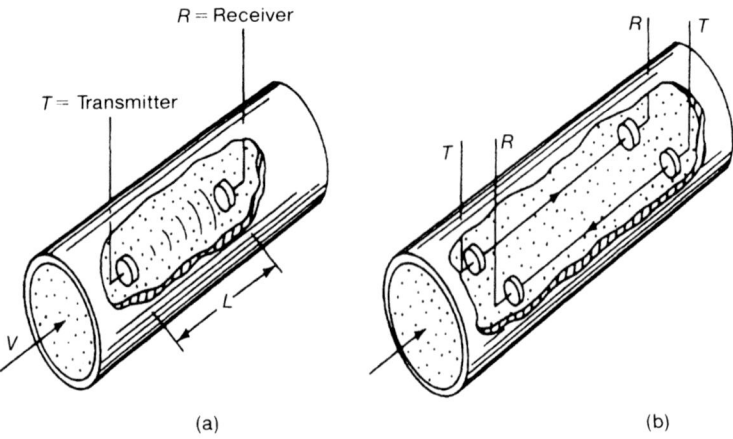

(a)

(b)

Fig. 8.9 Ultrasonic flowmeter.

Thus if L and c are known, measurement of Δt allows v_f to be calculated.

While L can usually be considered a constant, c varies with temperature for most fluids. Since c appears as c^2 in eqn (8.5), the error caused by a temperature change may be significant. Also, Δt is quite small, since v_f is normally a small fraction of c. For example, if v_f is 10 m/s, $L = 10$ cm and $c = 1500$ m/s, then $\Delta t \approx 0.4\,\mu s$, a short increment of time which can be hard to measure.

8.6 Vortex shedding flowmeters

The phenomenon of vortex shedding downstream from a bluff body immersed in a steady flow is well-known. It can arise unintentionally, as was the case at the Tacoma Narrows suspension bridge in the USA, which collapsed due to wind-induced oscillation in 1940 [4]. More usefully, vortex shedding is the basis of the vortex shedding flowmeter, shown in cross-section in Fig. 8.10. When the Reynolds number exceeds around 10 000 vortex shedding occurs reliably, and the shedding frequency f is

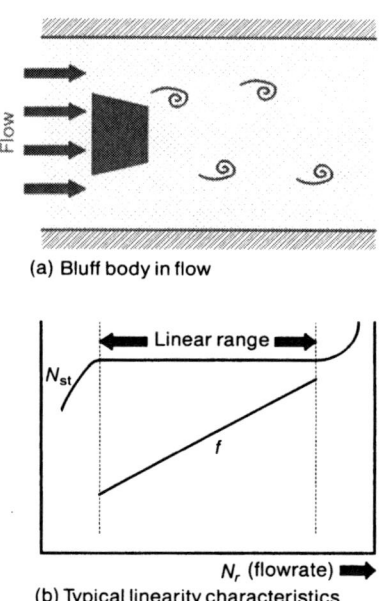

(a) Bluff body in flow

(b) Typical linearity characteristics

Fig. 8.10 Vortex shedding flowmeter.

given by eqn (8.6):

$$f = \frac{N_{st}V}{d}. \tag{8.6}$$

Where V is the fluid velocity, d the characteristic dimension of the shedding body, and N_{st} the Strouhal number, which is experimentally determined. By careful design N_{st} can be kept nearly constant over a large range of Reynold's number N_R and thus flow rate, making f proportional to V. Vortex shedding flowmeters are therefore pseudo-digital devices, in which the flow is measured by counting the vortex shedding rate (see Fig. 8.10(b)).

Various methods are employed to count the vortex shedding rate. The vortices cause local pressure changes (and hence force changes) on the bluff body, and piezoelectric or strain gauge transducers can be used to detect these. Hot-film transducers (see section 8.2.1) buried within the shedder can detect the local flow fluctuations caused by the vortices. Another common approach is to use the effect of the vortices on beams of ultrasound passed through the fluid. A mechanical approach is sometimes used, in which vortex-induced differential pressures cause oscillation of a small caged ball, the motion of which is detected by a magnetic proximity pickup.

References

[1] *Acoustics for Engineers*, J.D. Turner & A.J. Pretlove. Macmillan, 1991.
[2] *Instrumentation Reference Book*. B.E. Noltingk. Butterworths, 1990.
[3] *Measurement Systems*, E.O. Doebelin. McGraw-Hill, 1990.
[4] *Structures*, by J.E. Gordon. Penguin, 1978.

Signal conditioning circuits

9.1 Introduction

Most transducers produce an output which requires modification before it can be used. The signal amplitude may be inconveniently small: devices such as accelerometers for example typically have an output of only a few tens of pico-Coulombs. The signal may contain unwanted information, in the form of electronic noise. 'Mains hum' (50 Hz in Europe, 60 Hz in the USA) is a common problem, and filter circuits based on operational amplifiers ('op-amps') are frequently used to remove unwanted noise from signals. Although much signal processing is done digitally, most signals originate from analogue transducers, and the analogue output from such a sensor normally has to be conditioned (i.e. amplified and/or filtered) before it is in a form acceptable to an analogue-to-digital converter (ADC). Most initial signal processing is done using circuits based on op-amps, and it is with these devices and their applications that this chapter is concerned.

An amplifier is a device which increases the power contained within a signal. It is important to appreciate the difference between an amplifier, which provides a true power gain, and a device such as a transformer, which may increase the voltage in a signal, but gives no power gain. One way of visualizing a power gain is to think of it in terms of input and output impedances. If the power in a signal (as opposed to its voltage) is to be raised, then the amplifier must be capable of supplying the signal at an increased current as well as voltage. Thus, an amplifier must have a high input impedance, so that it only draws a little current, and a low output impedance, so that it can supply a larger current. In circuit analysis use is often made of the so-called 'ideal amplifier', which is assumed to have an infinite input impedance and a zero output impedance.

The 'ideal operational amplifier' is also assumed to have infinite gain, since this greatly simplifies the analysis of circuits containing these amplifiers. If the open-loop gain of the device is greater than about 1000, the gain of a closed loop circuit containing an amplifier will be almost independent of the open-loop gain. This can be shown by considering the ideal amplifier shown Fig. 9.1(a). The gain G of the amplifier V_2/V_1, where $V_{1,2}$ are phasors and G is in general complex. To keep things simple, the analysis which follows only considers real (i.e. DC voltage) inputs and

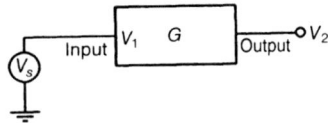

Fig. 9.1(a)

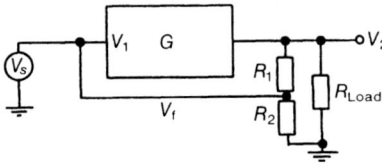

Fig. 9.1(b)

outputs. Nevertheless, the principle established holds true for the more general case of complex (i.e. alternating) signals.

Now examine the effect of adding feedback to the circuit, as shown in Fig. 9.1(b). The potential divider formed by R_1 and R_2 defines a fraction V_f of the output V_2 which is used for negative feedback. The Feedback Factor H is defined as:

$$H = \frac{V_f}{V_2}.$$

In general of course H is complex, although here:

$$H = \frac{R_2}{R_1 + R_2} = \frac{V_f}{V_2},$$

which is real. Now, the input V_1 is a combination of V_s and V_f, i.e. $V_s + V_f$. The output from the circuit is therefore:

$$V_2 = GV_1 = G(V_s + V_f) = G(V_s + HV_2).$$

The gain of the complete circuit is therefore V_2/V_s, where:

$$\frac{V_2}{V_s} = \frac{G}{(1 - GH)},$$

and if G is large:

$$\frac{V_2}{V_s} \approx -\frac{G}{GH} = -\frac{1}{H}.$$

i.e. the circuit gain depends only on the feedback factor H and not on the amplifier gain G, if G is large.

9.2 Operational amplifier circuit basics

The operational amplifier is conventionally represented in circuit diagrams by a triangle as shown Fig. 9.2. An op-amp is a DC coupled differential input device with a single output. This means that it amplifies the voltage difference between the two inputs. The DC gain of an op-amp is usually in the range 10^5–10^{10}. The device is designed so that its gain reduces at -6 dB/octave as frequency increases (see the discussion of gain-band-width products later).

The output signal voltage range is limited by the supply voltage. Generally the supply voltage is split, and ± 15 V is common which allows an output swing of about 28 or 29 V. Distortion resulting from an attempt to drive to output beyond these voltage limits is known as clipping.

The + and − symbols on Fig. 9.2 indicate the non-inverting and inverting inputs to the amplifier. When the noninverting (+) input is made more positive than the inverting (−) input the output becomes positive, and vice versa. The + and − symbols do not imply that one input must be kept more positive than the other, but indicate the phase the output signal has with respect to the input. For example, if a sinusoidal signal is connected to the inverting input of an op-amp, the output will be out of phase by $180°$ (π radians) with respect to the input.

The input impedance of an op-amp is the internal impedance between the inverting and non-inverting terminals. Typical values are between 1 MΩ and 100 MΩ. The output impedance of an op-amp is the internal impedance between the output and ground. Typical values are between 10 Ω–500 MΩ. The input-output impedance (the impedance between either input and the output) is very high, and can be assumed to be infinite. An op-amp may be represented for circuit modelling purposes by the equivalent circuit shown in Fig. 9.3. Note that as current is drawn from the device, the output voltage is reduced from its initial value V_0, and becomes:

$$V_{\text{out}} = V_0 - I_L Z_2 .$$

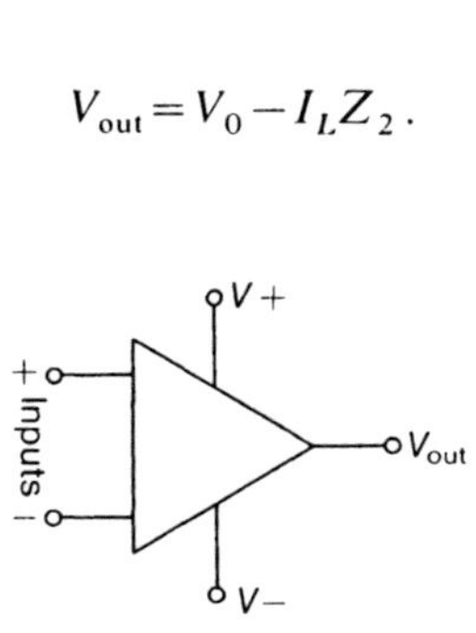

Fig. 9.2.

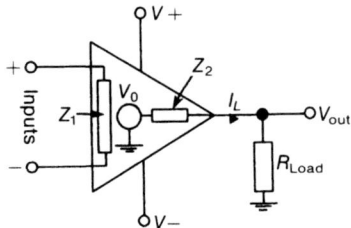

Fig. 9.3.

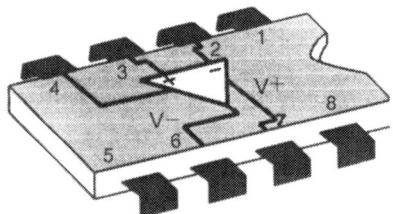

Fig. 9.4.

A common op-amp is the μA741C, or 741 for short. It comes in a variety of packages, but the most usual form is an 8-pin dual in-line (DIL) integrated circuit as shown in Fig. 9.4. The 741 is popular because it is simple to use, usually tolerates maltreatment without permanent damage, performs well and is cheap to obtain. However, it is worth remembering that much better op-amps can be obtained when necessary.

9.3 Analysing operational amplifier circuits

The behaviour of almost all op-amp circuits with external feedback can be understood by applying a pair of simple rules. These are:

I. It can be assumed (since the open-loop gain is so high) that there is no voltage difference between the two inputs.

II. The inputs may be assumed to draw no current.

Rule I needs some further explanation. It does not mean that an op-amp actually changes the voltage at one input in response to a change at the other. (If it did it would break rule II!). What it does mean is that the device adjusts its output so that the external feedback network brings the differential input voltage to zero if possible. If this is not possible, the output will approach one of the supply voltages (e.g. $+15$ or $-15\,\mathrm{V}$), and the resulting state is termed 'saturation'.

9.3.1 Example use of op-amp rules

To illustrate the use of these rules, consider the inverting amplifier circuit shown in Fig. 9.5. In an inverting amplifier the input signal is applied to the inverting ($-$) input, as shown in Fig. 9.5 and the output is out of phase by $180°$ (π radians) with respect to the input. Rules I and II can then be used to analyse the behaviour of this circuit as follows:

1. Point B is at ground voltage (0 V), so rule I implies that A is at ground also. Point A is often called a virtual earth.

2. Rule II implies that the current through R_1 is equal in magnitude and has opposite sign to the current through R_2, i.e.:

$$\frac{V_{in}}{R_1} = -\frac{V_{out}}{R_2}.$$

3. Therefore the gain of the circuit is A where:

$$\text{circuit gain } A = \frac{V_{out}}{V_{in}} = -\frac{R_2}{R_1}.$$

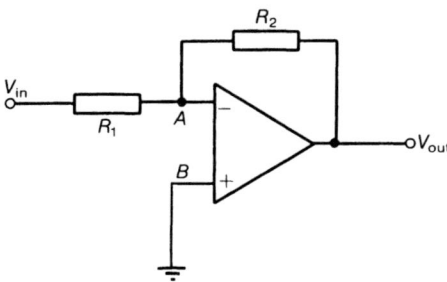

Fig. 9.5.

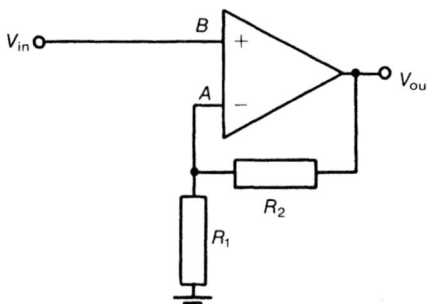

Fig. 9.6.

9.3.2 Example 2

We can use the op-amp rules similarly to calculate the gain of the non-inverting amplifier shown in Fig. 9.6. Rule I implies that $V_A = V_B$. A is the mid-point of a potential divider (as a consequence of rule II):

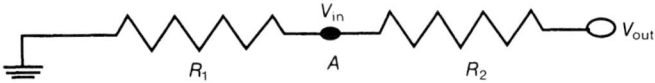

thus

$$V_A = \frac{R_1 V_{out}}{R_1 + R_2} \quad \text{and} \quad V_{in} = V_A,$$

so

$$\text{circuit gain } A = \frac{V_{out}}{V_{in}} = \frac{(R_1 + R_2)}{R_1} = 1 + \frac{R_2}{R_1}.$$

If we let R_1 become infinite and R_2 become zero, the equation above implies the gain is unity. The circuit is then:

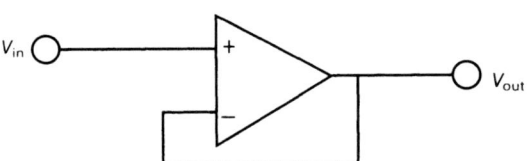

This is known as a *voltage follower*, and at first sight does not appear particularly useful. However, as well as posessing a gain of 1, the circuit has two other features: very high input impedance, and low output impedance. Thus, the circuit acts as a 'buffer', allowing current to be drawn from the output without affecting the input. This is often useful: for example, many sensors will not supply current without degrading the signal. The use of a voltage follower as a buffer between such a transducer and any subsequent circuitry avoids the problem.

9.3.3 Example 3

As well as resistors, operational amplifier circuits can include capacitors or inductors in their input circuitry or feedback loops. Circuits such as Fig. 9.7 can also be analysed by using the op-amp rules. This circuit has the useful property of integrating its input: thus, an acceleration record (e.g. from an accelerometer) can be converted into a velocity-time waveform. The circuit is similar to that of the inverting amplifier considered earlier, so point A is a virtual earth. By rule II therefore:

$$I_{\text{in}} = \frac{V_{\text{in}}}{R}.$$

Since the voltage at A is zero (a virtual earth), the feedback current I_f must be the capacitor charging current, i.e.:

$$I_f = C \cdot \frac{\text{d}}{\text{d}t}(V_{\text{out}}).$$

From rule II these currents must be equal in size and have opposite signs, i.e.:

$$\frac{V_{\text{in}}}{R} = -C \cdot \frac{\text{d}}{\text{d}t}(V_{\text{out}}),$$

i.e.

$$\text{d}V_{\text{out}} = -\frac{V_{\text{in}}}{(RC)}\text{d}t,$$

or

$$V_{\text{out}} = \frac{-1}{RC}\int V_{\text{in}} \cdot \text{d}t + \text{constant}.$$

Integrators do suffer from one problem: the output tends to increase at a steady rate ('ramping') until it saturates at the op-amp supply voltage.

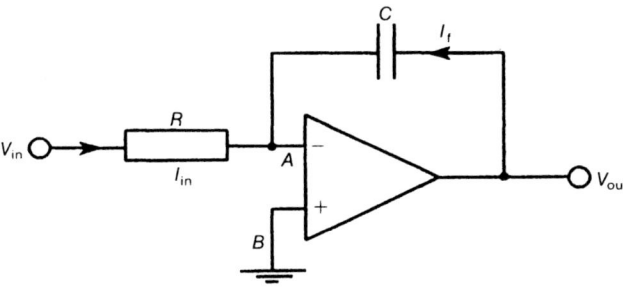

Fig. 9.7.

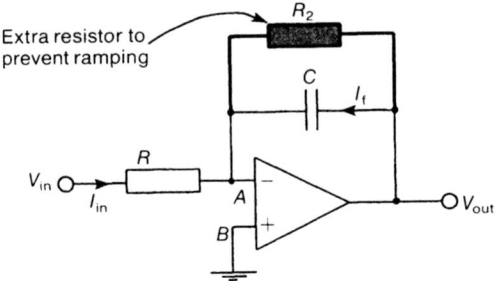

Fig. 9.8.

This effect occurs because there are almost always small residual DC voltages present in an op-amp circuit, and the integral of a DC voltage increases linearly with time. The solution to the problem is to modify the circuit so that the integrating action is lost at very low frequencies, which by definition includes DC. This is achieved by placing an extra resistor in parallel with the capacitor in the feedback loop, as shown in Fig. 9.8 The integrating action is thereby 'rolled off' below a frequency $f_{-3\,\mathrm{dB}}$, where:

$$f_{-3\,\mathrm{dB}} = \frac{1}{(2\pi R_2 C)} \text{ Hz}.$$

If the positions of the capacitor C and resistor R are exchanged in Fig. 9.7, the resulting circuit *differentiates* rather than integrates. By applying the op-amp rules (a good exercise!) it can easily be shown that for a differentiator:

$$V_{\mathrm{out}} = -RC\left(\frac{\mathrm{d}V_{\mathrm{in}}}{\mathrm{d}t}\right).$$

Differentiators are not so commonly used as integrators, because they suffer from a defect which has no easy solution: they increase the electronic noise level in a circuit. The reason for this is that, in general, the waveform of noise (for example, as viewed on an oscilloscope) is sharp and spiky, with many rapid changes of direction. These rapid alterations in gradient are of course magnified by the process of differentiation, which is essentially one of calculating rates-of-change. Thus, noise is magnified by a differentiator.

9.4 Frequency response and gain-bandwidth product

The preceding discussion has made no mention of frequency. It should be obvious that, since electronic devices are not perfect, no op-amp will function happily over an infinite range of frequencies. There is always a high-frequency limit, beyond which the circuit's output will become increasingly distorted. There is often a low-frequency limit as well: for example, the integrator discussed in section 9.3.3 was deliberately prevented from integrating low frequencies to avoid the problem of 'ramping'.

The high-frequency limit for an op-amp circuit can be predicted from the *gain-bandwidth product* (GBP), specified by the manufacturer. The GBP of an op-amp is the product of the DC gain and the frequency at which the gain is unity, as shown in Fig. 9.9. The rate at which gain decreases as the frequency increases is known as the roll-off. For all commercial op-amps the roll-off is -6 dB/octave. The GBP for a given device is a constant, and knowledge of its value allows the user to calculate the frequency range (or bandwidth) over which the gain remains approximately constant. For example, an op-amp having a GBP of 1×10^5 could be used to construct an amplifier circuit with a gain of 1 and a bandwidth of 100 kHz, or a circuit with a gain of 100 and a bandwidth of 1 kHz, or one with a gain of 1000 and a bandwidth of 100 Hz, etc. There is obviously a tradeoff between gain and bandwidth: the higher the gain, the smaller the usable bandwidth. If high gain *and* large bandwidth are required, the solution is to use a multi-stage amplifier design, in which the gain *at each stage* is limited to preserve bandwidth.

A diagram such as Fig. 9.9 in which the open-loop gain is plotted logarithmically against frequency is known as a gain-bandwidth diagram. An open-loop gain-bandwidth diagram can be used to calculate the bandwidth of a closed-loop circuit, since if the roll-off is -6 dB/octave the gain-bandwidth product is a constant. As we have seen, adding negative feedback to an amplifier reduces its gain, and since the gain-bandwidth product is constant, the bandwidth (the frequency range over which the gain is constant) must increase as the gain is reduced. Thus the gain-bandwidth diagram for an op-amp provides essential information, since

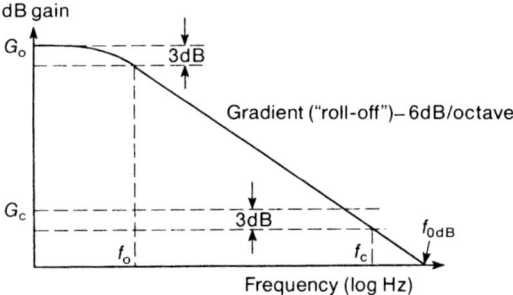

Fig. 9.9 Gain-bandwidth diagram. G_o is the open-loop gain, G_c the closed-loop gain. f_o and f_c are the -3 dB points. The gain-bandwidth product $G_o f_o = G_c f_c = (1 \times f_{o dB})$.

it allows the circuit designer to predict the frequency at which the gain of the circuit will begin to reduce.

9.5 The op-amp's departures from ideal behaviour

As mentioned in the introduction to this chapter, op-amps approach ideal characteristics (infinite gain, infinite input impedance, zero output impedance) with varying degrees of success. The purpose of this section is to outline some of the ways in which op-amps can misbehave. This misbehaviour largely arises because op-amp circuits are designed assuming ideal behaviour. When the op-amp refuses to oblige and obstinately persists in displaying 'real' characteristics, the designer can be caught out! The following sections contain descriptions of the main ways in which this non-ideal behaviour is measured and quantified, together with an indication of ways around the problems.

9.5.1 Common-mode interference

An ideal op-amp amplifies only differential signals, and has a gain of zero for common-mode signals (exactly identical signals which are applied to both inputs). Real amplifiers have non-zero values of common mode gain (CMG). The size of the CMG is a function of the magnitudes of the common and the differential input signals. A typical CMG value for a 741 op-amp with open-loop differential gain of 10^5 is about 3.

The relative magnitudes of the common-mode and differential gains is expressed in terms of a common mode rejection ratio (CMRR). CMRR is defined as the absolute value of the ratio of differential gain to common-mode gain:

$$CMRR = \frac{\text{Absolute value of open-loop differential gain } A(\omega)}{\text{Abs. value of CMG}}$$

i.e. $CMRR = |A(\omega)|/|CMG|$. Typical values are $|CMG = 5|$ and $|A(\omega)| = 10^5$, giving $CMRR = 20,000$ or $86\,dB$. The effect of CMG on V_{out} can be calculated as follows:

$$V_{out} = -(V_A - V_B)\cdot A(\omega) + V_{common\ mode}\cdot CMG$$

$$= -(V_A - V_B)\cdot A(\omega) + V_{common\ mode}\cdot \frac{A(\omega)}{CMRR}$$

$$= A(\omega)\cdot \left((V_A - V_B) + \frac{V_{common\ mode}}{CMRR} \right).$$

The CMRR is found from the device datasheet (usually in decibels, so it has to be converted into a ratio before use in circuit analysis). The datasheet also usually specifies an amplitude limit for common-mode inputs, beyond which the op-amp's output will become nonlinear.

9.5.2 Voltage supply rejection ratio (VSRR)

The characteristics of an op-amp are quoted on the manufacturer's datasheet for a standard supply voltage. If this voltage changes, in general the output of the amplifier will change. The size of this output variation is expressed in terms of the equivalent differential input voltage arising from a 1 V change in supply. In other words, a 1 V supply change will affect the output as though an *equivalent* input voltage were applied to the amplifier— and the size of the equivalent input producing the same effect as a 1 V supply change is quoted by the manufacturer as an indication of the device's sensitivity to power supply changes. The simplest way to explain how VSRR calculations are performed is by means of an example.

VSRR example calculation
Consider an inverting amplifier circuit such as Fig. 9.5. If the power supply voltage $(V+)-(V-)$ changes by 0.1 V, and the device has a VSRR of $-74\,dB$, by how much will the output change?

First, we have to convert the VSRR from decibels to a ratio:

$$\text{If } -74\,dB = 20 \cdot \log_{10}(x), \quad \text{what is } x?$$

$$x = 10^{-74/20}$$

$$= 2 \times 10^{-4} = 200\,\mu V/V.$$

So the op-amp's output reacts to a 1 V power supply change *as though a differential input of 200 μV were applied to the device*. Now, we want to know what the effect of a 0.1 V power supply change will be. To find out we need the op-amp's differential gain, which is obtained from the datasheet and which is frequency-dependant. The *maximum* differential gain is obtained at a frequency of 0 Hz (i.e. DC), so for a typical device with a differential gain of 10^5 at 0 Hz, the *maximum* voltage change at the output will be:

$$\Delta V_{out}(max) = 200(\mu V/V) \times 0.1(V) \times 10^5 = 2\,V.$$

9.5.3 Input offset voltage (IOV)

For an ideal op-amp V_{out} is zero when V_{in} is zero. Because components in symmetrical positions inside the amplifier are impossible to match precisely this rarely occurs in practice. A pair of terminals are provided as shown in Fig. 9.10 for balancing out the unwanted input offset. It should be borne in mind however that IOV is temperature sensitive, and balancing must therefore be carried out at the circuit's operating temperature.

9.5.4 Input bias current

It is necessary to supply small currents to both terminals of an op-amp to correctly bias the transistors within the amplifier. Earthing the unused input leads to an imbalance in these currents, and this results in an offset voltage appearing at the output of the device. To avoid this effect the unused terminal should be earthed through a resistor equal in value to the parallel combination of the input and feedback resistors. Figure 9.11 shows an inverting example. Biasing resistors are usually omitted from op-amp circuit diagrams for simplicity, but this does not mean that they are unnecessary!

9.5.5 Slew rate

The slew rate of an op-amp is the maximum possible rate of change of the output voltage with time. A value for slew rate is always specified on

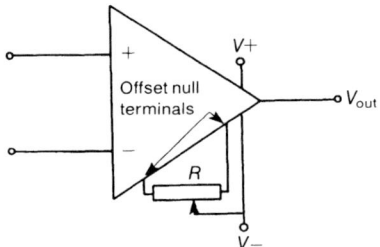

Fig. 9.10 Input offest nulling terminals.

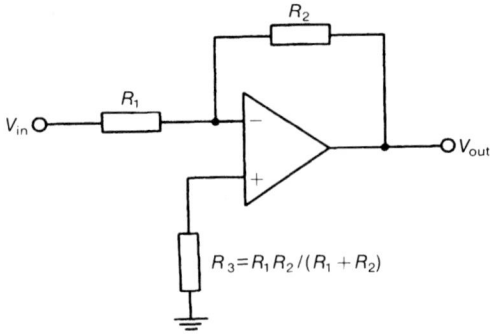

$$R_3 = R_1 R_2 / (R_1 + R_2)$$

Fig. 9.11.

the device's datasheet. Suppose the output from an op-amp is a sinusoid. If the output has the form:

$$V_{out} = V_0 \sin \omega t,$$

then the rate of change of the output will be:

$$\frac{dV_{out}}{dT} = \omega \cdot V_0 \cos(\omega t).$$

The *maximum* rate of change of V_{out} is obtained when the value of $\cos(\omega t) = 1$:

$$\left[\frac{dV_{out}}{dt} \right]_{max} = \omega V_0 = 2\pi f V_0,$$

where f is in Hertz and V_0 (the peak voltage) is in volts. Thus, to avoid output distortion it is necessary to choose an op-amp with a slew rate *greater* than $2\pi f V_0$. A typical slew rate is 0.5 V/μs (for a 741), but with special op-amps (such as the NE5539) it can be as high as 800 V/μs.

Care should be taken in applying the above analysis to non-sinusoidal periodic signals. To take a common example, suppose an op-amp is used to amplify a square wave of period T. It can be shown by Fourier analysis that a square wave consists of a sum of sinusoidal components at frequencies $1/T$, $3/T$, $5/T$, ... n/T, where the amplitude of each component is proportional to $1/n$. Thus a 1 kHz square wave contains sinusoidal components at frequencies of 1 kHz, 3 kHz, 5 kHz etc. The spectral content of a signal must therefore be considered before determining the slew rate required by a particular circuit.

9.5.6 Inverting and non-inverting circuit input impedances

Figures 9.5 and 9.6 show the circuits of inverting and non-inverting amplifiers. Clearly the input impedance of the non-inverting circuit is very high—and is equal to that of the op-amp itself. The inverting circuit however has a much lower input impedance. Since point A is a virtual earth, the input impedance of an inverting amplifier equals R_1.

9.6 Operational amplifier circuit selection

The earlier parts of this chapter provide a description of and guidance in the design and use of operational amplifiers. The following sections consist of a catalogue of a range of common op-amp circuits, which should be sufficient for most engineering purposes. The appropriate circuit can be selected by using the following tables. Table 9.1 contains circuits which convert a signal from one form to another. Table 9.2 contains a selection of circuits for conditioning voltage signals.

Table 9.1
Signal conversion circuits

To convert a signal from	To	Use circuits in Figs.
Current	Voltage	9.28
Voltage	Current	9.16
Charge	Voltage	9.29

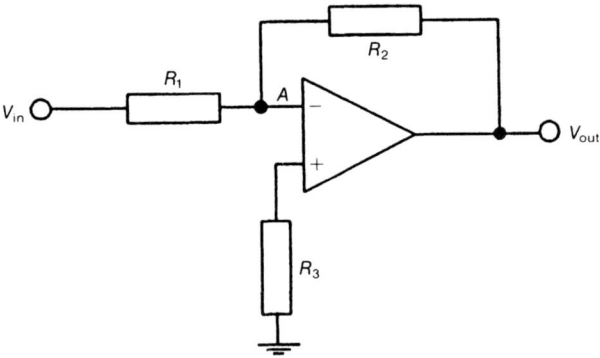

Fig. 9.12 Inverting amplifier.

9.7 Common operational amplifier circuits

9.7.1 The inverting amplifier

In an inverting amplifier the input signal is applied to the inverting ($-$) input as shown in Fig. 9.12, and the output is out of phase by $180°$ (π radians) with respect to the input. The gain of the circuit is:

$$\frac{V_{out}}{V_{in}} = \frac{-R_2}{R_1}.$$

The gain is unaffected by the presence of the bias resistor R_3. Since point A is at ground potential, the input impedance of an inverting amplifier is governed by R_1.

9.7.2 The non-inverting amplifier

A non-inverting amplifier is shown in Fig. 9.13(i). In this case the signal is applied to the non-inverting ($+$) input, and the output is in phase with the input. The gain is:

$$\frac{V_{out}}{V_{in}} = 1 + \frac{R_1}{R_2}.$$

Table 9.2
Circuits for conditioning voltage signals

Requirement	Solution
To amplify a voltage signal	All of it: And change phase, use circuit in Fig. 9.12 [NB reduced input impedance] — Without phase change, high input impedance, use circuit in Fig. 9.13. AC Part only: And change phase, use circuit in Fig. 9.14 — Without phase change, high impedance input, use circuit in Fig. 9.15
To attenuate a signal	Use an inverting amplifier (circuits in Fig. 9.12 or 9.14) since minimum gain for non-inverting circuits is 1. Use two inverting circuits in series for a non-inverting attenuator
To add two signals and amplify the sum	Use a summing ampifier, circuit in Fig. 9.17
To subtract two signals and amplify their difference	Use a differential amplifier, circuit in Fig. 9.18
To pass only some frequencies from a signal (i.e. filtering action)	All frequencies below a specified cut-off, use circuit in Fig. 9.19. All frequencies above a specified cut-off, use circuit in Fig. 9.20. All frequencies within a specified band, use circuit in Fig. 9.21. All frequencies below a specified band, use circuit in Fig. 9.21 (but also see Fig. 9.22)
To precision rectify (convert AC to DC)	Use circuits in Fig. 9.23 or 9.24
To integrate a signal with respect to time	Use circuit in Fig. 9.25
To differentiate a signal with respect to time	Use circuit in Fig. 9.26
To compare two signals, and indicate which is largest	Use circuit in Fig. 9.27.

Circuits in Figs. 9.19–9.21 are first order filters. See reference [1] if filters with a higher specification (in particular, greater roll-off) are required.

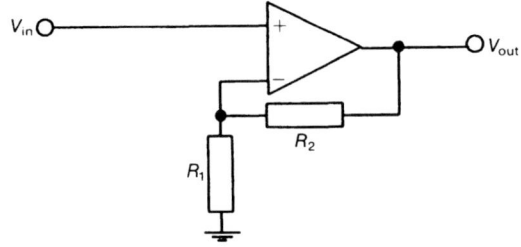

Fig. 9.13(i) Non-inverting amplifier.

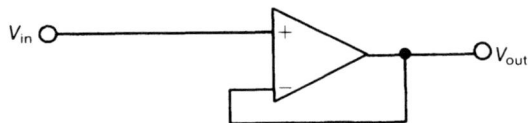

Fig. 9.13(ii) Follower (unity gain amplifier).

Note that if $R_2 = 0$ and R_1 is infinite a unity-gain voltage follower results as shown in Fig. 9.13(ii). Almost no current is drawn by the input of a non-inverting amplifier, and so no load is applied to the circuit or device providing V_{in}. A non-inverting amplifier consequently protects a signal source from any loads applied at V_{out} because of the very high input impedance of the op-amp. This should be contrasted with the behaviour of the inverting amplifier, where current is drawn from V_{in} at the input and flows through resistors R_1 and R_2.

It should also be noted that a non-inverting amplifier may not be used as an attenuator, i.e. it cannot have a gain of less than unity. An inverting amplifier may have a gain which is less than one.

9.7.3 AC amplifiers

If only AC signals are to be amplified it is good practice to 'roll off' the gain at frequencies close to DC. This prevents the amplifier becoming saturated by DC offset voltages. An inverting AC amplifier is shown in Fig. 9.14. The gain of the amplifier is:

$$\text{Gain} = \frac{V_{out}}{V_{in}} = -\frac{(\text{feedback impedance } Z_2)}{(\text{input impedance } Z_1)}$$

$$= -\frac{2\pi j f R_2 C}{1 + 2\pi j f R_1 C}.$$

When $f = 0$ Hz (i.e. a DC signal), the gain $V_{out}/V_{in} = 0$. As $F \to \infty$, $V_{out}/V_{in} \to -R_2/R_1$: the same as the gain of an ordinary (i.e. non-frequency

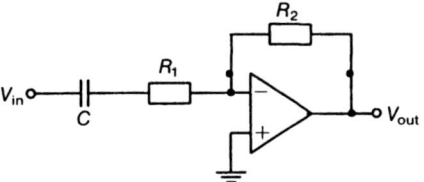

Fig. 9.14 Inverting AC amplifier.

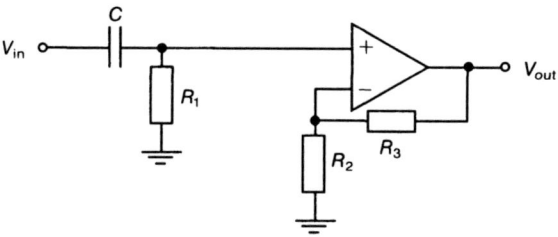

Fig. 9.15 Non-inverting AC amplifier.

dependent) inverting amplifier (circuit 9.12). The half-power ($-3\,\text{dB}$) point is at a frequency $f_{-3\,\text{dB}}$ such that:

$$f_{-3\,\text{dB}} = \frac{1}{2\pi R_1 C}.$$

At frequencies well above $f_{-3\,\text{dB}}$ Hz the capacitor may be thought of as essentially a short circuit, and the input and output impedances are the same as those of a DC coupled inverting amplifier. Figure 9.15 is a non-inverting AC amplifier. The gain is:

$$\frac{V_{\text{out}}}{V_{\text{in}}} = \left(\frac{1}{1 + 1/(2\pi j f R_1 C)}\right) \cdot \left(1 + \frac{R_3}{R_2}\right).$$

At $0\,\text{Hz}$ the gain is zero. At high frequencies (well above $f_{-3\,\text{dB}} = 1/[2\pi R_1 C]$) the gain tends towards:

$$\frac{V_{\text{out}}}{V_{\text{in}}} = 1 + \frac{R_2}{R_1},$$

which is the same as the gain of a simple non-inverting amplifier (circuit 9.13(i)).

9.7.4 The constant current source, or voltage/current converter

An op amp current source such as Fig. 9.16 gives near ideal behaviour without the voltage offset inherent in transistor current sources. The

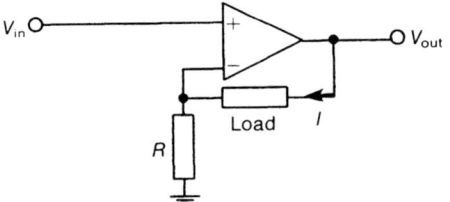

Fig. 9.16 Constant current source (current/voltage converter).

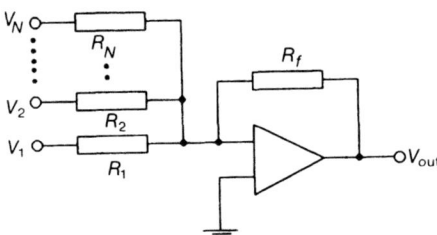

Fig. 9.17 Weighted-sum amplifier.

current I through the load is V_{in}/R, which is independent of any changes in the load impedance. Voltage V_{in} is usually derived from a potential divider, or for greater stability a zener diode may be used. The circuit can also be used to convert a signal from voltage to current form.

9.7.5 Summing amplifiers

The output of Fig. 9.17 is a sum of the applied voltages V_1, V_2 etc. If the resistors R_1, R_2 etc. have different values the output is a weighted sum, given by:

$$\frac{V_{out}}{R_f} = -\left(\frac{V_1}{R_1} + \frac{V_2}{R_2} + \cdots + \frac{V_N}{R_N}\right).$$

The circuit may be extended to have as many inputs as required.

9.7.6 Differential amplifiers

Figure 9.18 shows a differential amplifier. The most useful property of a differential amplifier is that it rejects common-mode signals originating at the signal source. The output is a weighted function of the input signals:

$$V_{out} = \frac{R_4}{R_3}\left(\frac{1+R_2/R_1}{1+R_4/R_3}\right)\cdot V_{s2} - \left(\frac{R_1}{R_2}\right)\cdot V_{s1}.$$

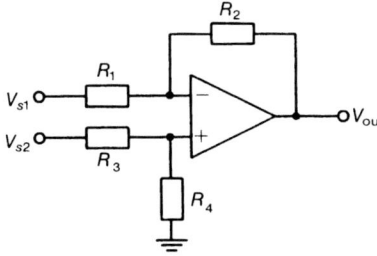

Fig. 9.18 Weighted differential amplifier.

The differential amplifier rejects signals appearing at its input terminals if the signals are common to both inputs, i.e. if they have a common mode. The output in the presence of a common-mode voltage at the input terminals of the op-amp is:

$$V_{out} = -\left\{ \frac{R_2}{R_1} \cdot V_s + \left(1 + \frac{R_2}{R_1}\right) \cdot \left(\frac{V_{common}}{CMRR}\right) \right\}$$

(where $V_s = V_{s1} - V_{s2}$ and CMRR = common mode rejection ratio). For a common-mode voltage at the signal source the output voltage is:

$$V_{out} = \frac{-R_2}{R_1} \cdot \left\{ V_s + \frac{V_{common}}{CMRR} \right\}.$$

To maximise the rejection of any common mode signals applied to V_{s1} and V_{s2}, component values should ideally be chosen such that:

$$\frac{R_1}{R_2} = \frac{R_3}{R_4} \quad \text{and} \quad R_1 = R_3 + R_4.$$

9.7.7 Filter circuits

The simplest kind of op-amp filter contains one capacitor for each cut-off frequency, i.e. one capacitor for low and high pass, and two for bandpass and bandstop filters. This type is called a first-order filter (since its behaviour is described by a first-order differential equation). The next degree of complexity is obtained by a second order filter (which contains two capacitors in its high- and lowpass forms, and four for a bandpass system), and so on. The higher the order of the filter, the closer it approaches ideal characteristics, in that the roll-off or attenuation outside the passband is steeper. As a rule of thumb, the attenuation outside the passband of a filter in decibels (dB) per octave will be six times the filter order. Thus, a first order filter will have a roll-off of $-6\,\text{dB/octave}$, a second order filter will have a roll-off of $-12\,\text{dB/octave}$, and so on. Filters higher than fourth order are seldom constructed using op-amps, due to the complexity of the circuits involved. If very high performance filters are needed the best solution is to use special-purpose filter devices. Filter design is a specialised topic, and although first order filters such as those described here are adequate for many purposes, there will inevitably be occasions when something better is needed. Reference [1] contains further details of filter design.

The non-ideal characteristics of any op-amp mean that bias currents and gain effects must be considered in practical filter circuit designs. A DC analysis of bias currents must be undertaken, and the unused input must be earthed through an appropriate resistor as discussed in section 9.5.4.

It must also be remembered that the gain of any op-amp circuit using feedback decreases above a frequency determined by the gain-bandwidth product as discussed in section 9.4.

9.7.7.1 Low-pass filter

Figure 9.19 is a first order low-pass filter. The $-3\,\mathrm{dB}$ cut-off frequency is given by:

$$f_H = \frac{1}{2\pi R_2 C_2}.$$

The DC gain is R_2/R_1, and the roll-off in the stop band is $-6\,\mathrm{dB/octave}$. The following example illustrates the filter design process:

Example calculation
Design a first-order low-pass filter with $f_H = 2\,\mathrm{kHz}$ and DC gain of 5. We require:

$$1/(2\pi R_2 C_2) = 2000 \quad \text{and} \quad R_2/R_1 = 5.$$

We have two equations with three unknowns. We can therefore choose one of the three components arbitrarily. For example, let $C_2 = 0.01\,\mu\mathrm{F}$. Then from the equations above, $R_2 = 7.96\,\mathrm{k\Omega}$ and $R_1 = 1.59\,\mathrm{k\Omega}$. The final step

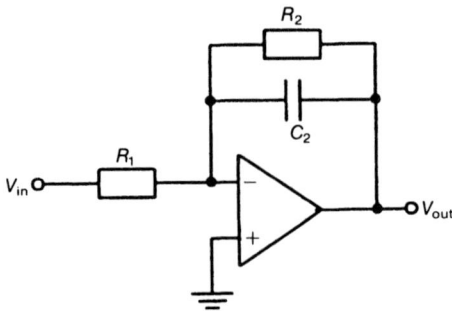

Fig. 9.19 First-order low-pass filter.

is to calculate a suitable value for the bias resistor: assuming the filter is to be used well inside its passband, R_3 should be given a value equivalent to that of the parallel combination of R_1 and R_2, i.e.:

$$R_3 = \frac{R_1 R_2}{R_1 + R_2} = \frac{7960 \times 1590}{7960 + 1590} = 1325 \, \Omega.$$

9.7.7.2 High-pass filter

Figure 9.20 is a first-order high pass filter. The $-3 \, \text{dB}$ cutoff frequency is:

$$f_L = \frac{1}{2\pi R_1 C_1},$$

and the high frequency gain (i.e. well beyond the cut-off frequency) is $-R_2/R_1$. The roll-off below f_L is $-6 \, \text{dB/octave}$.

9.7.7.3 Band-pass filter

Figure 9.21 is a first order bandpass filter. The low and high cut-off frequencies are respectively:

$$f_L = \frac{1}{2\pi R_1 C_1} \quad \text{and} \quad f_H = \frac{1}{2\pi R_2 C_2}.$$

The mid-frequency gain (passband gain) is $-R_2/R_1$. The roll-off outside the passband is $-6 \, \text{dB/octave}$.

Example
Design a filter with $f_L = 2 \, \text{kHz}$, $f_H = 5 \, \text{kHz}$, and a passband gain of 10. The design equations are used as follows. We have:

$$2000 = 1/(2\pi R_1 C_1) \quad \text{and} \quad 5000 = 1/(2\pi R_2 C_2)$$

and

$$-R_2/R_1 = 10$$

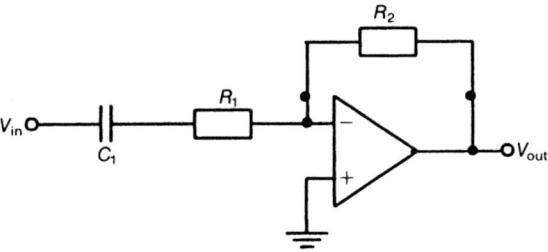

Fig. 9.20 First-order high-pass filter.

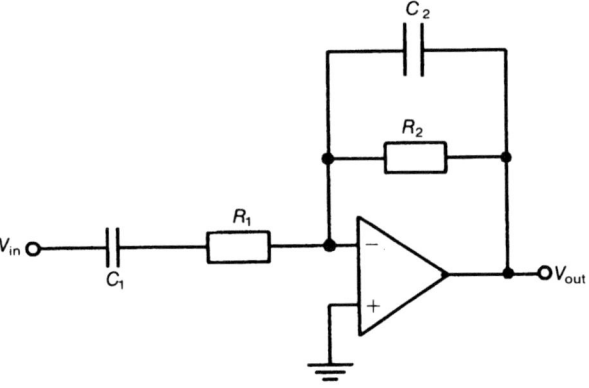

Fig. 9.21 First-order band-pass/band-stop filter.

i.e. three equations and four unknown variables. As before, we choose a value (using realistic component sizes!) for one of the four arbitrarily. For example, let $C_1 = 0.01\,\mu\text{F}$. Then:

$$R_1 = 7.96\,\text{k}\Omega, \quad R_2 = 796\,\Omega, \quad \text{and} \quad C_2 = 0.04\,\mu\text{F}.$$

The design is concluded by fixing a value for the bias current equalization resistor R_3, such that:

$$R_3 = \frac{R_1 R_2}{R_1 + R_2} = \frac{7960 \times 796}{7960 + 796} = 724\,\Omega.$$

9.7.7.4 Bandstop filter (notch filter)

A bandstop filter is essentially the same as a bandpass filter—the only difference is (as shown in Fig. 9.22) the positions of f_L and f_H are reversed. Instead of the *passband* being between f_L and f_H, the *stopband* lies between

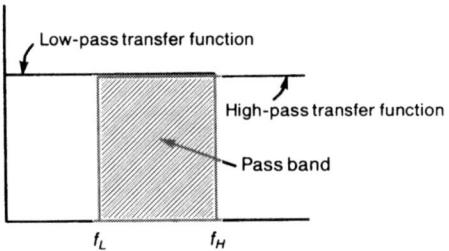

Fig. 9.22(a) Band-pass filter transfer function.

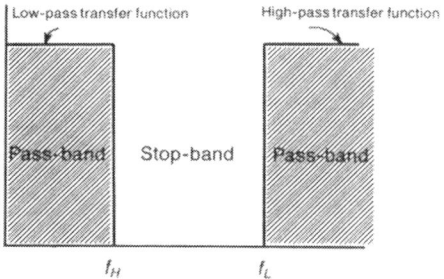

Fig. 9.22(b) Band-stop filter transfer function.

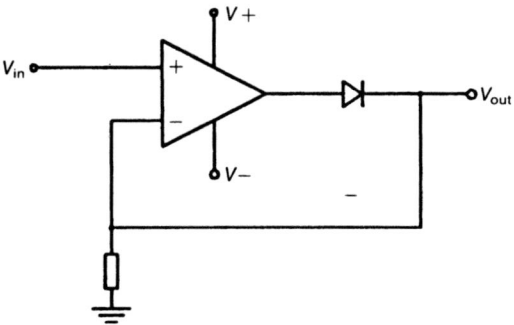

Fig. 9.23 Precision rectifier.

them, and there are two passbands as shown. The design equations are the same as those used for bandpass filter design.

9.7.7.5 The precision rectifier

A diode used as a half-wave rectifier is of limited application in low voltage (less than about 1 V) circuits, because of the voltage drop which occurs across a diode during forward conduction. With a silicon diode, at least 0.7 V has to be applied in the forward direction before any output voltage appears. Figure 9.23 avoids this problem. When the input is positive, the high gain of the op-amp causes the diode to conduct when the non-inverting input is only a few microvolts more positive than the inverting input.

Figure 9.23 suffers from two limitations. First, during negative input excursions a large differential voltage is applied to the op-amp, driving it into saturation. An op-amp may take up to 50 μs to recover from saturation, and this places limits on the high frequency performance of the circuit. Second, an op-amp which can tolerate large differential inputs without damage must be used. These can be expensive.

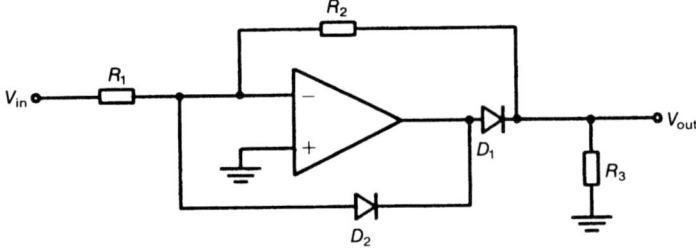

Fig. 9.24 Practical precision rectifier.

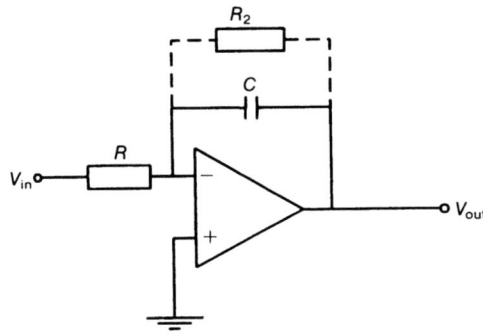

Fig. 9.25 Integrator.

The arrangement shown in Fig. 9.24 avoids these difficulties. Diode D2 prevents the op-amp becoming saturated, and a much faster response is achieved.

9.7.7.6 The integrator

An op-amp can be used to make an almost perfect integrator. Figure 9.25 shows the arrangement used. As shown previously in section 9.3.3, the output voltage is found from:

$$V_{out} = \frac{-1}{RC} \int V_{in} \cdot dt + \text{constant}.$$

The constant is determined by the initial conditions, and can appear as a DC voltage added to the output signal.

An integrator is a useful circuit if a 90° phase shift is required, since the integral of a cosine function is a sine wave. This is often used to test the purity of a sinusoid—the signal and its integral are used to form a Lissajous figure on an oscilloscope, and if the gain of the integration circuit is unity, a perfect circle results from a pure sinusoid. Integrators can also be used to generate saw-tooth waveforms, since if a square wave is integrated, a saw-tooth results.

One problem with an integrator of the type shown in Fig. 9.25 is that its output tends to 'ramp' or increase steadily until it saturates at the supply voltage, due to op-amp offsets or unmatched bias currents. For this reason many op-amp integrators are zeroed periodically by closing a switch (often a FET) placed across the capacitor, so that only the drift over a short time period matters. An alternative solution is to place a large value resistor R_2 in parallel with C, which has the effect of rolling-off the integrating action at frequencies below $f_{-3\,\mathrm{dB}}$ where $f_{-3\,\mathrm{dB}} = 1/(2\pi R_2 C)$.

9.7.7.7 The differentiator

If the resistor and capacitor used in the integrator (Fig. 9.25) are interchanged as shown in Fig. 9.26, a differentiating circuit results. The output is given by:

$$V_{\mathrm{out}} = -RC \frac{\mathrm{d}V_{\mathrm{in}}}{\mathrm{d}t}.$$

Differentiators can be awkward to use as they tend to amplify any noise present at the input to the circuit. This is because noise usually takes the form of a rapidly-changing or 'spiky' waveform, and the differential of anything changing quickly is likely to be large. In addition, differentiators can suffer from instability problems at high frequencies. Both these shortcomings can be overcome to some extent by rolling-off the differentiating action at high frequencies. This is achieved by adding R_x and C_x to the circuit as shown.

9.7.7.8 The comparator

A comparator is simply an open-loop op-amp, (see Fig. 9.27), which is driven into positive or negative saturation according to the difference between the two voltages at its inputs. Since the open loop gain of an op-amp is so high (typically $>10^5$ or 100 dB for a 741), the input voltages have to be equal to within a small fraction of a millivolt for the device's

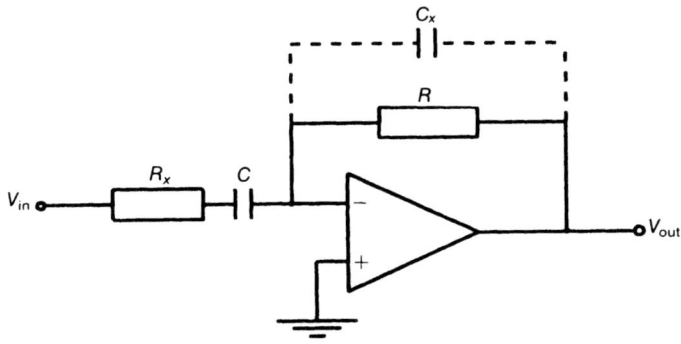

Fig. 9.26 The differentiator.

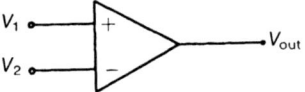

Fig. 9.27 The voltage comparator.

output not to be saturated. The polarity of the saturated output indicates the direction of the inequality relating the input voltages. Although ordinary op-amps can be used as comparators, a device such as the 741 takes an appreciable time to recover from saturation. Its use as a high-speed comparator will also be limited by the slew rate, (typically 10^6 V/s for a 741). It is much better to use a special op-amp which is designed to act as a comparator. Examples include the LM306, LM311 and LM393, all of which are produced by National Semiconductor. These devices have a very fast recovery time and slew rates in excess of 10^9 V/s.

There are three points to note about circuit design using comparators:

1. Since there is no feedback, the inputs are not necessarily at the same voltage.

2. Again because there is no feedback, the input impedance is not necessarily constant. As a result the input signal sees a changing load and a changing (small) input current as the comparator output switches.

3. Some comparators will only permit limited differential inputs, as small as 5 V in some cases. Check the device datasheet before deciding to use it in a design!

9.7.7.9 Current to voltage converter

The simplest current to voltage converter is, of course, a resistor. However, the trouble with using resistors is that they present a non-zero impedance to the source of input current. This can cause difficulties if the current source has very little compliance. (A constant current source can only maintain a constant current through a load over a finite range of load voltage. The output voltage range over which a current source is well behaved is known as its compliance). Photovoltaic cells for example have a very small compliance. Figure 9.28 shows how to use an op-amp as a current-to-voltage converter. The inverting input is a virtual earth. The output voltage is determined by the feedback resistor; in Fig. 9.28 it is 1 V/µA of input current. The resistor between the non-inverting input and ground ensures that the input bias currents match.

9.7.7.10 The charge amplifier

A charge amplifier is usually used with capacitive sensors, and for piezo-electric sensors which act as a charge source. A charge amplifier is only

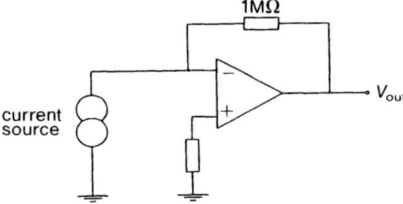

Fig. 9.28 Current-voltage converter.

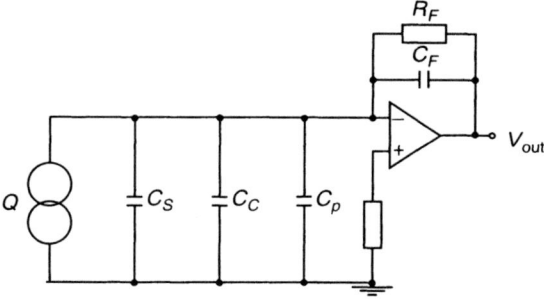

Fig. 9.29 Charge amplifier.

sensitive to variations in charge, which means that almost any length of cable can be used to connect a sensor to a charge amplifier without affecting the sensitivity. As shown in Fig. 9.29, a charge amplifier consists of an op-amp with capacitive feedback. This gives the circuit an input capacitance C_p, where:

$$C_p = C_F(G-1) \quad \text{(where } G \text{ is the open loop gain of the op-amp).}$$

The output voltage is given by:

$$V_{out} = \frac{Q \cdot G}{C_S + C_C - C_F(G-1)}.$$

Since the open loop gain of an op-amp is very high, the output voltage becomes:

$$V_{out} = \frac{Q}{C_F}.$$

With C_F constant, the output voltage V_{out} is directly proportional to the input charge Q. The cable capacitance C_C normally has a negligible effect.

Only when the cable is so long that the size of C_C approaches C_F will the sensitivity of the circuit be affected. The feedback resistor R_F is there to provide a suitable input bias current, and should be chosen so that:

$$R_F \geqslant \frac{1}{2\pi f C_F},$$

where f is the operating frequency.

References

[1] The *Art of Electronics*, by Horowitz and Hill. Cambridge University Press.

Signal conversion and data acquisition

10.1 Introduction

Measurements taken over time form a 'time history'. This time history may be analogue, such as the output from an accelerometer or may be inherently digital, such as the output from an optical shaft encoder, in which the signal is one of a finite number of discrete states. As digital data is highly immune to corruption and suitable for computer-based processing, many analogue signals must be transformed into a digital, or sampled form, before processing, even those which are continuously measured in time.

The transformation of continuous data to digital data is known as 'analogue to digital (A/D) conversion' and the opposite process as 'digital to analogue (D/A) conversion'.

10.2 Analogue to digital conversion

It can be helpful to think of A/D conversion as consisting of two processes: sampling and quantization. When the continuous signal is sampled, a 'snapshot' is taken of the signal in time. This sample, which will probably be a voltage, must then be quantized: translated into a binary number.

While A/D converters do not generally use such distinct processes it is a useful way of picturing the process as it highlights the two types of approximation inherent in the conversion. The first is an approximation in time, in that a continuous signal is sampled at discrete intervals, and this, as we will see later, limits the frequency range of that data. The second approximation is due to forcing a signal with an infinite number of possible values into one with a limited number of binary digits. This is a quantization error and restricts the dynamic range of the signal after conversion.

The quantization error caused by A/D conversion can be illustrated by a simple example. An analogue signal $v(t)$ which varies from 0 to 6 V is to be converted to a 4-bit digital signal. The number of different output states into which the signal is divided is given by $2^4 = 16$ and the relationship between the analogue input and the digital output is shown in Fig. 10.1. The effect of quantization can be seen in Fig. 10.2 which shows

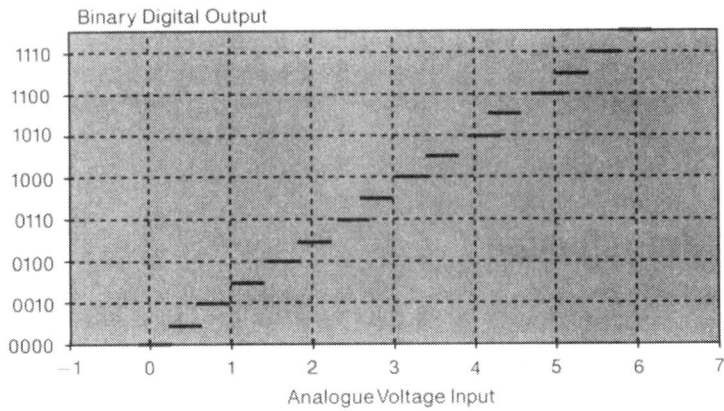

Fig. 10.1 Analogue to Digital conversion.

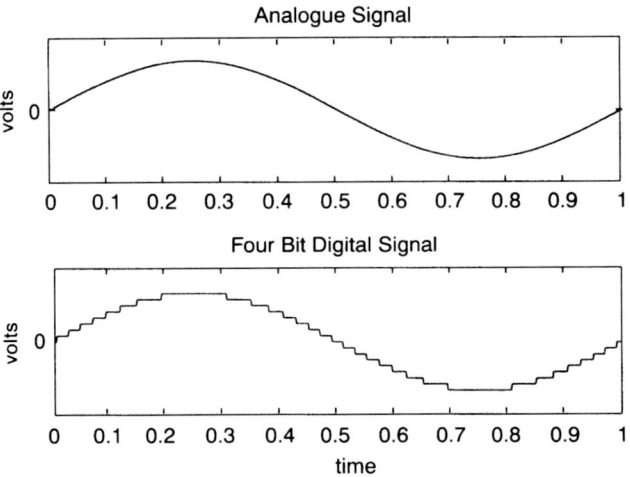

Fig. 10.2 Analogue signal and its four bit binary equivalent.

an analogue signal and a four bit digital conversion of that signal (a four bit signal allows 16 different values).

Suitable choice of, and correct use of modern equipment should make this quantization error small, and once a signal is in digital form it is highly immune to noise and degradation. However, there is one practical point which must be borne in mind to avoid serious problems with quantization errors, and this is correct matching of the analogue range

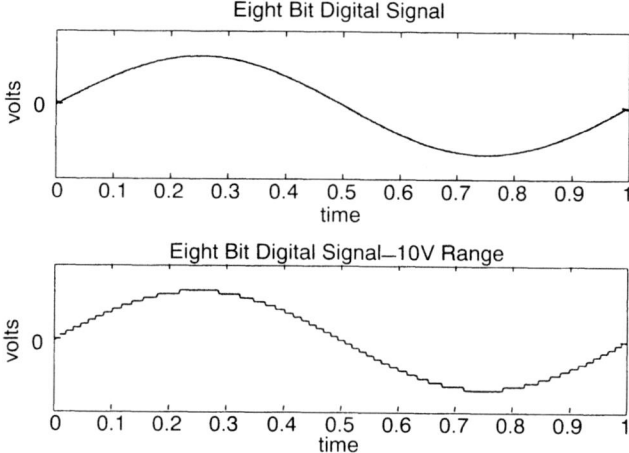

Fig. 10.3 Eight-bit digitization and incorrect use of range.

to the input range of the A/D Converter. The first graph in Fig. 10.3
shows that even an eight-bit digitization (allowing 256 discrete values)
provides a good approximation to the signal. Note that in this case, unlike
Fig. 10.1, the data may be positive or negative so one of the bits acts as
a sign bit. The second graph in Fig. 10.3, however, shows the same signal,
also digitized using eight bits, but without using the full range of the A/D
converter. The A/D converter here is set to ± 10 V, while the signal itself
is only ± 1 V. This demonstrates the importance of considering quantization errors, especially in signals with a wide dynamic range.

10.2.1 Quantization errors

From Fig. 10.1 it can be seen that a single digitized number is associated
with a range of analogue values, any of which could have been quantized
to that number. The width of a 'step' in Fig. 10.1 represents the range of
analogue values which could potentially lead to each digital output and
this will be equal to the full analogue input range of the device, divided
by the number of discrete digital states. The device represented in
Fig. 10.1 has a nominal input range of 0–6 V, and the step width will be
$(6.0 + 2(0.2))/16 = 0.4$. The '0.2s' are due to the fact that we have chosen
to place the maximum and minimum analogue values in the middle of a
digital step. The uncertainty caused by this quantization means, for
example, that the digitized number 0101 could have been caused by any

value between 1.8 and 2.2, which may be thought of as a possible error of magnitude $\pm \delta/2$, where δ is the analogue value equating to the least significant bit in the digital representation. For most signals, this error can be assumed to be a random process uncorrelated with the signal [1] which will have a uniform probability distribution, $p(e)$, about the true value (Fig. 10.4). The number of bits required in an A/D converter is determined by the dynamic range (the ratio, in dBs, between the largest and smallest amplitudes) which has to be measured. This, in turn, is governed by the signal to noise ratio (SNR), where:

$$\text{SNR} = 10 \log_{10}\left(\frac{\sigma_x^2}{\sigma_e^2}\right), \qquad (10.1)$$

and where σ_x^2 is the variance of the signal to be measured and σ_e^2 is the variance of the error, which may be written as [see, for example reference 2 p. 462]:

$$\sigma_e^2 = \int_{-\infty}^{+\infty}\left((e-\bar{e})^2 p(e)\right)de = \int_{-\delta/2}^{\delta/2}\left(\frac{e^2}{\delta}\right)de = \frac{\delta^2}{12}. \qquad (10.2)$$

The relationship between the variance of the signal to be measured and the number of bits used is a trade-off and will need careful consideration if an application is close to the limit of the equipment to be used. The greater the variance the higher the SNR due to quantization but also the higher the likelihood of clipping. To ensure a very low probability of clipping in a random signal we may choose to ensure that the maximum range of a system with b bits, given by $2^b\delta$, is greater than

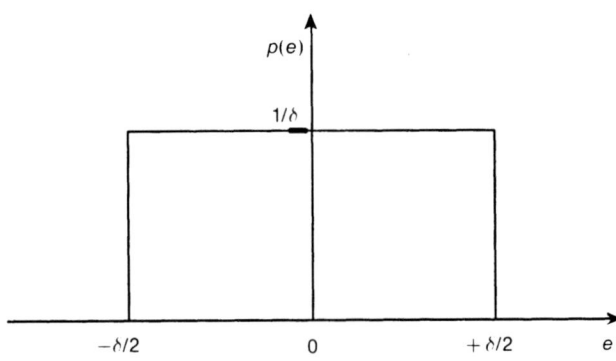

Fig. 10.4 Probability distribution of quantization error.

five times the rms value of the signal:

$$\sigma_x \geqslant \frac{2^b \delta}{5}. \tag{10.3}$$

Substituting into 10.1 gives:

$$\text{SNR} = 10 \log_{10}\left(\frac{12(2^{2b})}{5^2}\right)$$

$$= 10b \log_{10}(4) + 10 \log_{10}\left(\frac{12}{5^2}\right)$$

$$\approx 6b - 3.2 \, \text{dB}.$$

Hence in our earlier example, when our signal was coded into three bits plus a sign bit, the SNR would be about 15 dB. If we were to use more of the available range and simply ensure that the maximum range is greater than three times the rms of the signal the SNR would be extended to 19 dB but there would be increased clipping. Table 10.1 shows the SNRs which can be expected from different systems using the more conservative factor of the range being 5 times the rms signal.

Other errors which may be present in an A/D converter include:

Linearity—this is usually specified by the manufacturer of the A/D device as the maximum deviation from a straight line drawn between the full scale output and zero.
Gain error—most devices have a nulling input to deal with this effect. However, it is often found that the null setting required is temperature dependent, and there is no easy solution to this problem.
Offset error—similar to the above. Once again temperature dependence can be a problem.

Exaggerated examples of these types of error can be seen in Fig. 10.5.

Table 10.1
SNRs for different numbers of bits with the range set to 5 times the rms signal amplitude

Number of bits	SNR (dB)
8	45
12	69
16	93
24	140

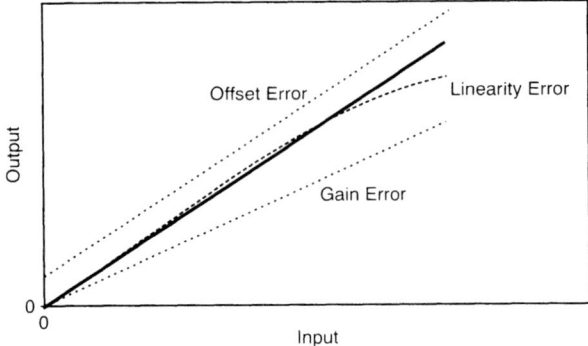

Fig. 10.5 Different error types which may be encountered in A/D conversion.

10.3 Computer-based data acquisition

In many general-purpose measurement situations, it is most convenient simply to use a purpose built card or external adapter designed to allow a computer or workstation to acquire data directly. In such circumstances the user rarely needs to worry about the details of the A/D conversion method used. The questions which the user needs to concentrate on instead include cost and 'user-friendliness' along with the following:

- *Sampling rate.* The sampling rate at each input channel should be well over twice the maximum frequency contained in the signal to be sampled at that channel. For very high speed data acquisition an oscilloscope (perhaps able to 'download' data to a computer) or high speed, stand alone data logger may be a sensible option. Issues associated with sampling are dealt with in more depth in chapter 11.

- *Signal type.* Current, charge, voltage, and impedance. Usually these cards are designed to present a relatively high impedance to a voltage source. It may be necessary to obtain or build a buffer or amplifier.

- *Signal amplitude.* The range of the input signal will have to be matched to that of the card. Some cards have switchable ranges (-1 to $1\,\mathrm{V}$, -5 to $+5\,\mathrm{V}$ and 0 to $10\,\mathrm{V}$) for example. In other circumstances an additional amplification stage may need to be included.

- *Signal dynamic range.* The dynamic range of the signal determines the number of bits required in the data encoding. In some cases a form of data compression may also be required to get the most out of a given acquisition system.

- *Triggering.* If the signal to be measured is a transient one, the user may wish to start acquiring the data automatically. This can be achieved using a trigger which begins the data acquisition when a particular condition, such as a threshold voltage on the input line, has occurred. There are commonly two types of trigger, a software trigger and a hardware trigger. A hardware trigger uses a pin on the data acquisition card as an 'external trigger'. When that trigger pin sees a high input it begins to read in data from the input line, or lines. It is then up to the user to construct a circuit which feeds the correct signal to the trigger pin at the correct time. The circuit in Fig. 10.6 uses a buffer amplifier, a voltage comparator, a monostable and a second comparator to produce a negative leading edge TTL trigger signal when the input voltage rises above a certain level.

  The problem with a trigger such as this when measuring transient signals, is that by the time the computer begins to log the data the early part of the transient has been lost. Turning down the trigger level on the first comparator will help this, but will also make the trigger more susceptible to noise. An alternative is to use a card which allows internal, or software triggering. In this case when the computer is not acquiring data, it nonetheless reads data into a buffer in memory continuously. It examines each data point and compares it to a pre-set level. If the input remains below that level then it simply looses the earliest piece of data in the buffer and looks at a new data point. Once a new data point exceeds the trigger level, however, it begins to store the buffered information in memory and will continue to do so until the end of the acquisition period. In this way, the computer can use 'pre-trigger' and log data from before the triggering event took place.

- *Signal length.* The amount of data you are able to acquire in one go will be limited by the amount of computer memory and, at lower sample rates, disk space available on your computer.

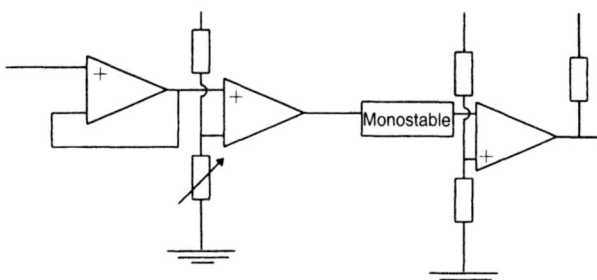

Fig. 10.6 Triggering circuit for hardware trigger.

- *Environment.* A conventional desktop computer, fitted with a data acquisition card, may well be an appropriate cost-effective way of logging data in the laboratory. If, however, such a task must be carried out in the open air, or over a long period in an industrial environment, for example a more rugged, specialist data logger or industrial PC may be appropriate.

References

[1] Oppenheim, A.V. & Schafer, R.W. *Digital Signal Processing.* Prentice Hall, 1975.
[2] Spencer *et al.*, *Engineering Mathematics*, Volume 1, Van Nostrand, UK, 1977.

Signal analysis—frequency domain techniques

11

11.1 Introduction

Up to this point, this book has discussed methods for measuring physical parameters, such as temperature, length, or acceleration, converting those measurements into an electrical signal, modifying or filtering that signal, and acquiring the signal on a computer. The next stage in the measurement process is to analyse that data.

In some cases, data analysis is relatively simple and can be performed by studying the change in value of the measurement over time—the time history. There are often occasions, however, when the time history is too complex to allow much to be gleaned from it without further processing.

Probably the most important single tool used in signal processing is the Fourier Transform. The Fourier Transform enables a time history to be broken down into its frequency components. This often reveals far more about the structure and origin of the signal than the raw time data alone. As an example, Fig. 11.1 shows a noisy accelerometer signal displayed in

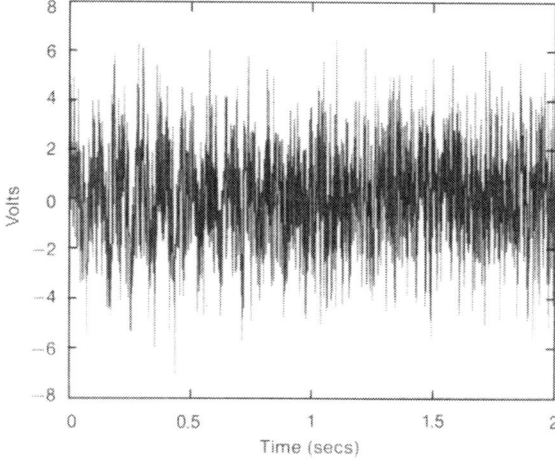

Fig. 11.1 Noisy signal measured in time.

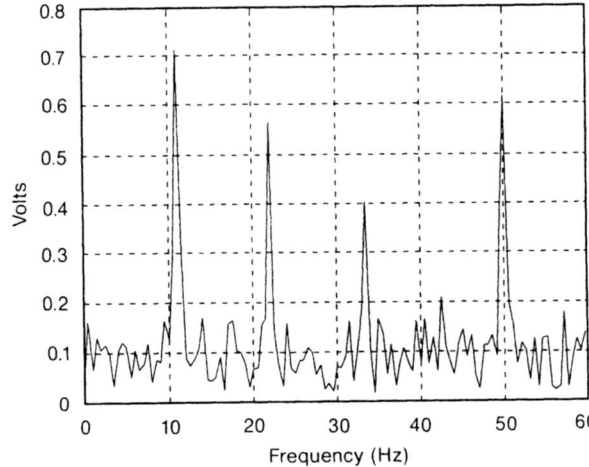

Fig. 11.2 Frequency spectrum of same signal.

the time domain. In the time domain it is nearly impossible to draw meaningful conclusions about the nature of this signal.

Figure 11.2 shows a frequency spectrum of the same signal. It is now clear that the signal consists of a strong component at about 11 Hz, followed by weaker components at about 22 and 33 Hz. It is likely that whatever caused the 11 Hz component also produced the harmonics (integer frequency multiples) at 22 and 33 Hz. There is also a strong component at around 50 Hz, which might be attributed to mains hum, depending on the measurement. The noise, which obscured details in the time history, has a broadband frequency characteristic and so is 'spread around' at a relatively low level in the frequency spectrum. This allows relatively weak features, such as the 33 Hz harmonic, to be observed.

In addition to allowing the examination of signals as frequency spectra, the Fourier Transform is an essential pre-cursor to many signal-processing techniques that are more advanced. There are many texts available which cover the details of frequency analysis [1,2] and its mathematical derivation. This chapter simply reviews some of the basic points of theory and refers the interested reader to these texts for a fuller picture.

11.2 The Fourier series

The essence of frequency analysis is the fact that a broad class of signals can be broken down into sums of simple individual components, and further, that the response of a broad class of systems to a signal can be deduced from the response of that system to these simple components.

Fourier analysis breaks down the time domain signal into sets of sine waves. The process can best be understood by examining the Fourier series representation of periodic signals. A periodic signal is one that repeats itself after a fixed time-period, such as the one shown in Fig. 11.3. The time taken before the signal repeats itself is known as the period of that signal. Of course, most signals are not periodic, but the simple theory applicable to such signals can be generalized.

Fourier proved that most periodic functions $f(t)$ of period T may be represented as a set of sinusoids:

$$f(t) = \frac{a_0}{2} + \sum_{n=1}^{\infty} \left\{ a_n \cos\left(\frac{2\pi nt}{T}\right) + b_n \sin\left(\frac{2\pi nt}{T}\right) \right\}. \tag{11.1}$$

This means that a periodic signal can be formed by adding together sets of sine and cosine harmonics (signals having integer frequency relationships with the fundamental). While the sum is infinite, a good convergence can usually be seen after the summation of just a few terms. All that is required is to calculate the values of a_0, a_n, and b_n. These can be found using the following relationships in which the period under consideration starts at time t_1 and ends at time t_2 (i.e. $T = t_2 - t_1$):

$$a_0 = \frac{2}{T} \int_{t_1}^{t_2} f(t) \, dt, \tag{11.2}$$

$$a_n = \frac{2}{T} \int_{t_1}^{t_2} f(t) \cos\left(\frac{2\pi nt}{T}\right) dt, \tag{11.3}$$

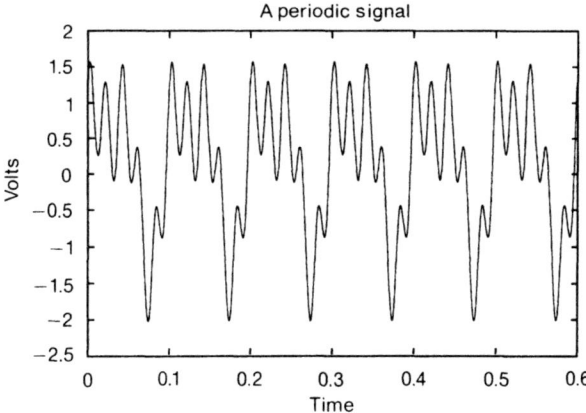

Fig. 11.3 Example of a periodic signal.

$$b_n = \frac{2}{T} \int_{t_1}^{t_2} f(t) \sin\left(\frac{2\pi nt}{T}\right) dt. \tag{11.4}$$

Derivations of these relationships may be found in references [1 and 2] and most textbooks on engineering mathematics. To see how this works in practice we will construct a square wave from a set of sinusoids.

We wish to form a signal such as the one shown in Fig. 11.4. The solution for a_0, a_n, and b_n is found using eqns (11.2), (11.3) and (11.4):

$$a_0 = \int_0^1 dt - \int_1^2 dt = 0,$$

$$a_n = \int_0^1 \cos(n\pi t) \, dt - \int_1^2 \cos(n\pi t) \, dt$$

$$= \left[\frac{1}{n\pi} \sin(n\pi t)\right]_0^1 - \left[\frac{1}{n\pi} \sin(n\pi t)\right]_1^2$$

$$= 0,$$

$$b_n = \int_0^1 \sin(n\pi t) \, dt - \int_1^2 \sin(n\pi t) \, dt$$

$$= -\left[\frac{1}{n\pi} \cos(n\pi t)\right]_0^1 + \left[\frac{1}{n\pi} \cos(n\pi t)\right]_1^2$$

$$= \frac{1}{n\pi} \{-(-1)^n + 1 + 1 - (-1)^n\}$$

$$= \begin{cases} 0 & \text{for } n \text{ even} \\ 4/n\pi & \text{for } n \text{ odd.} \end{cases}$$

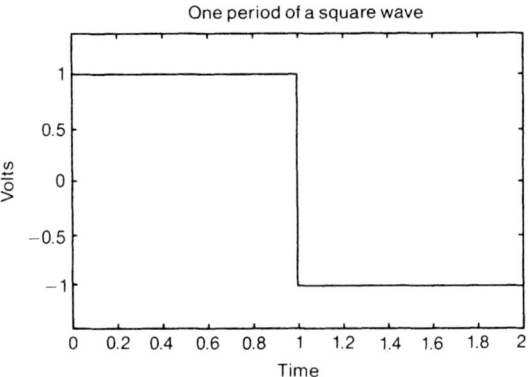

One period of a square wave

Fig. 11.4 Square wave.

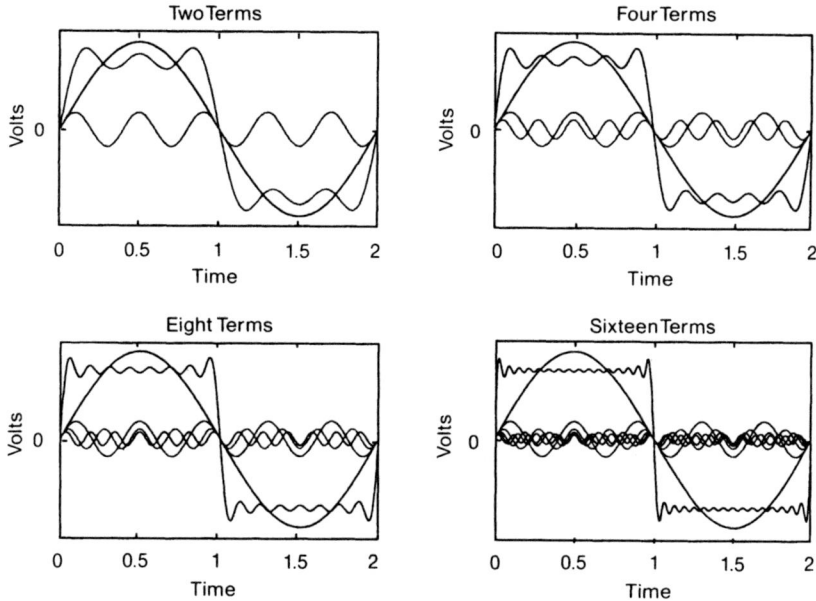

Fig. 11.5 Building a square wave by adding harmonics of a sine wave.

Hence for this square wave, all the 'a' terms are zero, as are the even 'b' terms. The odd 'b' terms decrease in amplitude as n increases. The result of adding the series term by term is shown in Fig. 11.5. It can be seen that there is an overshoot that remains at the discontinuities.

Such a series can be represented as a spectrum in the frequency domain as well as in the time domain. A frequency domain representation of the first eight terms in this spectrum is shown in Fig. 11.6. A striking illustration of how this sum forms a square wave is provided by a mesh plot. The effect of adding further terms can be seen in the sum of the first 30 terms shown in Fig. 11.7. Those familiar with complex notation for oscillations may also recognise the complex version of the series which may be written as:

$$f(t) = \sum_{n=1}^{\infty} F_n e^{j2\pi nt/T}. \tag{11.5}$$

11.3 The Fourier transform

The Fourier series is limited in that it can only represent periodic functions. The frequency content of other signals such as transients may be examined using an extension to this theory known as the Fourier transform. If the time function is not periodic then it must be formed by

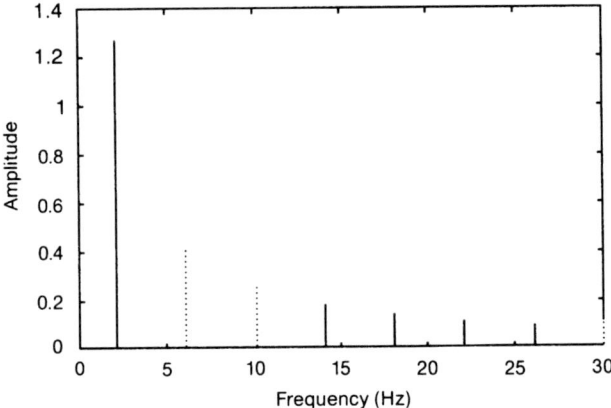

Fig. 11.6 The first eight terms in the spectrum of a square wave.

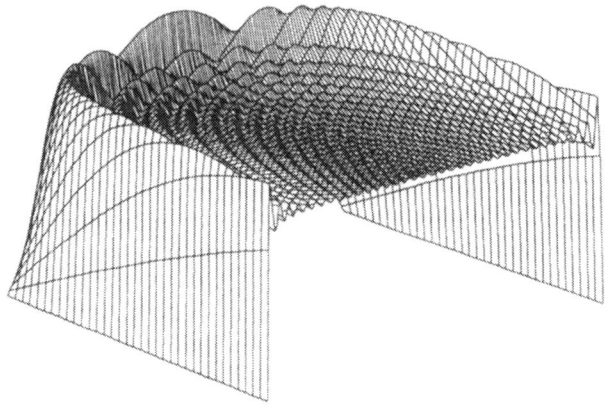

Fig. 11.7 Sum of the first 30 terms of the Fourier series.

integrating across a continuous frequency spectrum rather than summing individual values in a discrete spectrum. The frequency spectrum $S(f)$ of a time history $x(t)$ can (in most circumstances) be written as:

$$S(f) = \int_{-\infty}^{\infty} x(t)\, e^{-j2\pi f t}\, dt, \qquad (11.6)$$

and the time history calculated from the spectrum using the inverse transform:

$$x(t) = \int_{-\infty}^{\infty} S(f)\, e^{j2\pi f t}\, df. \qquad (11.7)$$

11.4 Digital frequency analysis—the discrete Fourier transform

The Fourier transform pair (eqns (11.6) and (11.7)) is not suitable for implementation on a digital computer as it stands. Data must first be sampled to create a digital signal rather than a continuous signal. This process leads to an important phenomenon known as aliasing. This section describes some of the effects and dangers of the digitization and processing.

The Fourier transform of eqn (11.6) turns a continuous time signal into a continuous frequency spectrum. The transform pair, which transforms a time history sampled with a sampling interval of Δt into a spectrum, is given by:

$$S(f) = \sum_{n=-\infty}^{\infty} x(t_n) e^{-j2\pi f t_n} \tag{11.8}$$

$$x(t_n) = \frac{1}{f_s} \int_{-f_s/2}^{f_s/2} S(f) e^{j2\pi f t_n} \, df. \tag{11.9}$$

Where $f_s = 1/\Delta t$ is the sampling frequency and $t_n = n\Delta t$ is the time after n samples. Equation (11.8) is still not suitable for use in practice as it involves an infinite summation, while a practical spectral analysis can only work on a finite number of samples. If we limit the number of data points considered to N then the record length T will be $N\Delta t$. This is equivalent to truncating the summation and has several important effects on the transformation. The result is only valid over a limited frequency range and the equations act as if both the time history and the spectrum repeat themselves outside of their limits. This imposes a periodicity on both sets of data, which needs careful treatment to avoid errors. The resulting spectrum is also defined only at discrete frequency points rather than a continuous spectrum. The resulting transform pair is known as the discrete Fourier transform (DFT) and is defined as:

$$S(k) = \frac{1}{N} \sum_{n=0}^{N-1} x(n) e^{-j2\pi kn/N} \tag{11.10}$$

$$x(n) = \sum_{k=0}^{N-1} S(k) e^{j2\pi kn/N}. \tag{11.11}$$

The DFT algorithm is now straightforward to implement on a digital computer. As it stands, however, it is relatively inefficient and calculations are computationally very demanding. Most practical analysers use one of several much faster versions of the DFT known as fast Fourier transform (FFT) algorithms. These rely on a symmetry in the calculations which exists if N is a power of 2, so an FFT can only be implemented on data sets of length 256, 512, 1024 etc. data points.

11.5 Sampling and aliasing

Perhaps the most important, and, certainly from the user's point of view, the most dangerous result of sampling the time signal is aliasing. Aliasing causes signals with a frequency of more than a half of the sampling frequency ($f_s/2$) to be interpreted as a frequency below $f_s/2$.

As an example, Fig. 11.8 shows a set of data sampled at 20 Hz and Fig. 11.9 shows the 6 Hz continuous signal which *may* have been sampled to produce that data set. Whatever signal actually did produce it, an FFT algorithm would interpret it as a 6 Hz signal and a digital to analogue converter (DAC) would recreate it as a 6 Hz signal. However, Fig. 11.10 shows that the sampled sequence could also have been created by a 14 Hz continuous signal.

In fact, any frequencies in the series 6, 14, 26, 34 ..., could have led to the same sampled data sequence. The relationship between these frequencies, all of which can cause 'aliasing' at 6 Hz, can be seen from Fig. 11.11. The frequencies causing aliasing at 6 Hz with a sampling frequency of 20 Hz are 20 ± 6 Hz, 40 ± 6 Hz etc. The principal of aliasing is identical to the use of a stroboscope to examine rotating machinery and vibrating structures. In this case, the stroboscope may be set at a 'sampling

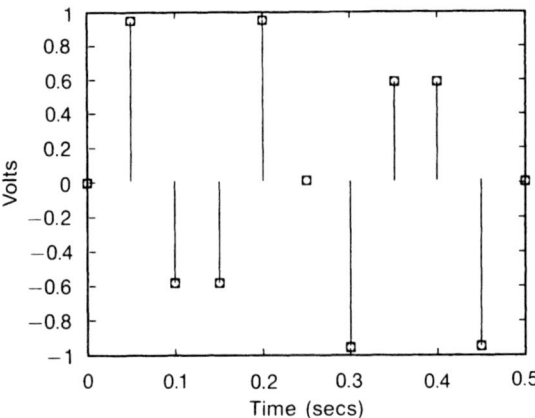

Fig. 11.8 Unknown signal sampled at 20 Hz.

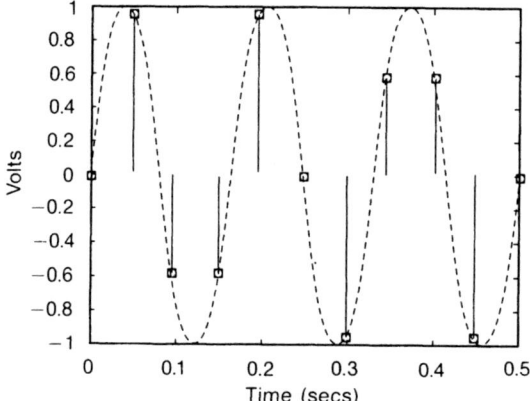

Fig. 11.9 Six Hz signal sampled at 20 Hz.

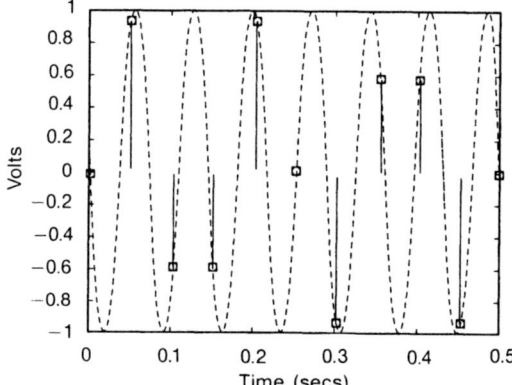

Fig. 11.10 Fourteen Hz signal sampled at 20 Hz.

frequency' which has the vibration frequency as a harmonic (an integer multiple). The structure then appears stationary (i.e. creates an alias frequency of zero). The 'sampling frequency' is then adjusted away from this point so that the structure appears to move at a very low frequency, allowing it to be observed easily. This low frequency is an aliasing frequency and is equal to the difference between the vibration frequency and the nearest harmonic of the stroboscope frequency. To link this with

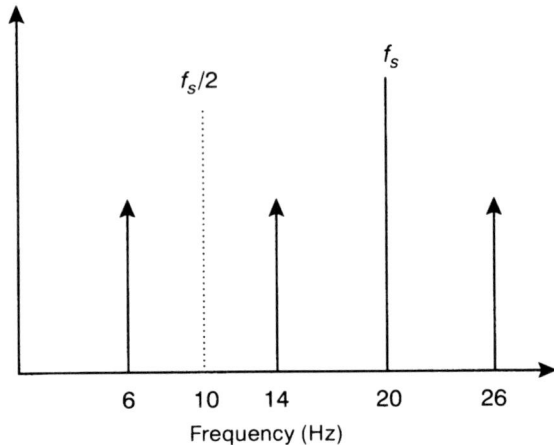

Fig. 11.11 Spectrum of alias frequencies.

Fig. 11.11, our structure might be vibrating at 26 Hz and we could set our stroboscope to flash (sample) at 26 Hz as well, creating an alias of 26–26 = 0 Hz, thus making the structure appear stationary. If we then adjusted our strobe to sample at 20 Hz it would appear to move at 26–20 = 6 Hz just as the alias signal in Fig. 11.11 does.

When we sample data, therefore, the presence of any signals at frequencies above $f_s/2$ will give spurious results when we look at the data in the frequency domain. This frequency of one half of the sample rate is known as the Nyquist frequency and for reliable measurements to be taken there must be no components of the signal at a frequency higher than the Nyquist frequency. This means that signals that are to be analysed by the FFT technique must first be passed through a low pass filter known as an anti-alias filter. Given that no filter is perfect, a typical strategy might be to choose a filter with a cut-off just above the highest frequency in which you are interested and then sample the signal with a sample rate of four to five times that frequency. As a useful rule of thumb choose the cut-off frequency f_c to be a little above the frequency range you wish to study and then set the sampling frequency f_s using the equation:

$$f_s \geqslant 2 \times 10^{0.3A/B} \times f_c,\qquad(11.12)$$

where: A = dynamic range required (dB);
B = slope of anti-alias filter (dB/Octave).

11.6 Windowing

We have seen that aliasing is the result of sampling data in the time domain. The second approximation introduced by the DFT of eqns (11.10) and (11.11) was to consider only N samples rather than summing to infinity. This causes the DFT algorithm to assume that the time history repeats itself outside of the range considered. To see how this works in practice, Fig. 11.12 shows a 4.5 Hz signal and the result in the time domain of sampling it for just one second.

When this signal is analysed, however, the DFT assumes that it is repeated and effectively periodic. This may produce large discontinuities in the signal, as shown in Fig. 11.13

The effect of these discontinuities can be seen clearly in a frequency spectrum. A Fourier transform of a sine wave should produce a single spike in the frequency spectrum, such as those shown in Fig. 11.6.

However, the FFT of this sample, shown in Fig. 11.14, is far less sharp than this and suggests that energy has been spread out (or 'leaked' out)

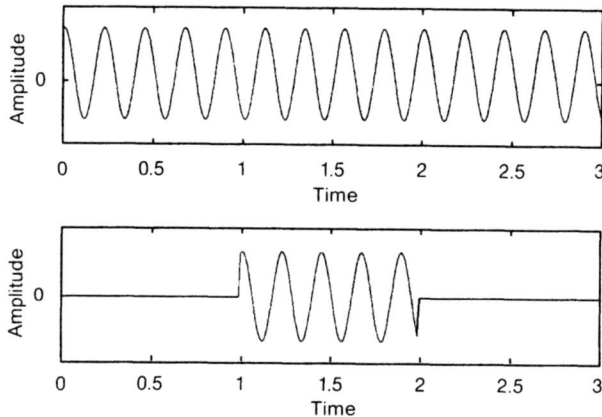

Fig. 11.12 Continuous signal and signal sampled for just one second.

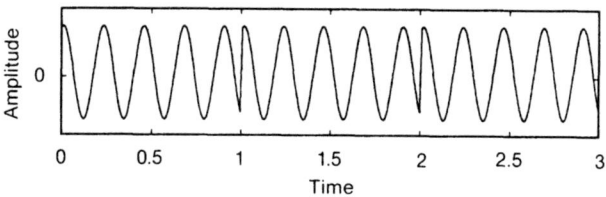

Fig. 11.13 Truncated signal as seen by DFT.

in the frequency domain away from the 4.5 Hz peak. This frequency leakage is due to the discontinuities created the sudden end of the sampled time domain data. A signal that starts from zero and has decayed to zero by the end of the sample would not display the same discontinuities and would not, therefore, suffer from the frequency leakage. A signal that does not behave like this, however, may be forced to do so by multiplication by a function that reduces its amplitude at the start and finish of the sampled sequence. Such a function is known as a *window* and multiplication in the time domain by such a function is known as *windowing*.

To see how this might work in practice, consider a time signal containing a frequency of about 32 Hz and a second frequency of 33 Hz at a much lower amplitude. Figure 11.15 shows about half a second of this signal in the time domain. The true spectrum of the continuous signal is shown in Fig. 11.16.

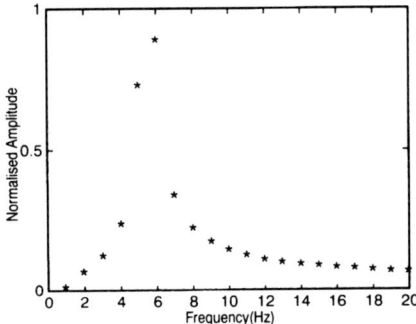

Fig. 11.14 FFT of data from Fig. 11.12.

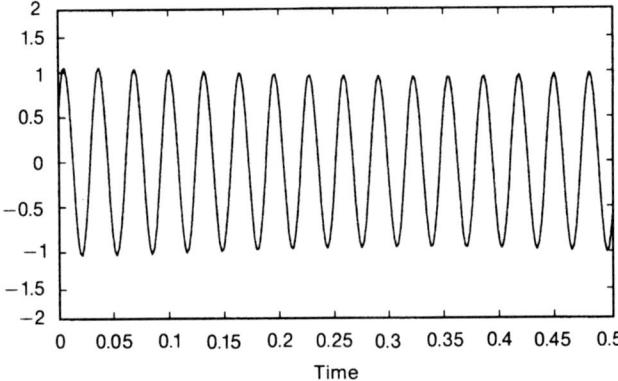

Fig. 11.15 Signal consisting of 32 Hz component and a lower amplitude 33 Hz component.

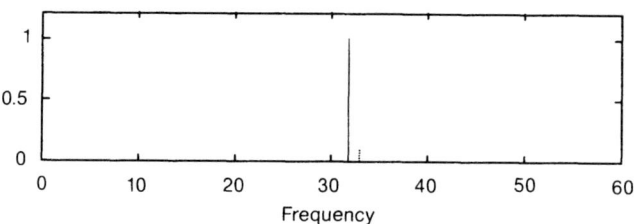

Fig. 11.16 True spectrum of signal shown in Fig. 11.15.

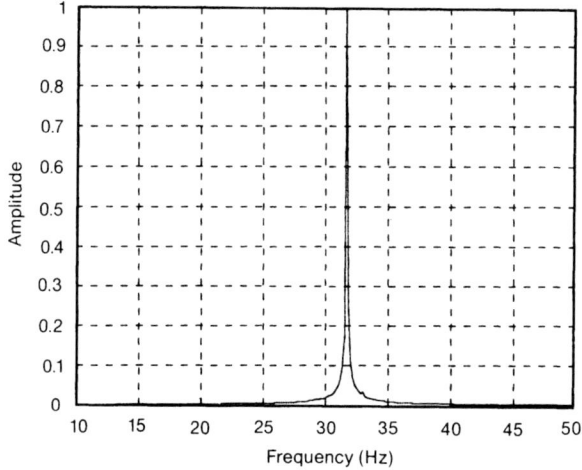

Fig. 11.17 FFT of 3 s, truncated sample without window.

If we simply take a three-second sample of this signal and perform an FFT on it, the result is rather different and Fig. 11.17 shows how frequency leakage from the larger peak has completely obscured the smaller peak.

To reduce this leakage and ensure that the signal starts and ends at zero we multiply the time sequence by a window. A common window is the Hanning window, based on a cosine function, although there are many different windows in use with a variety of properties. Figure 11.18 shows the time series both before and after the Hanning window has been applied.

Figure 11.19 shows an FFT of the signal taken after the window has been applied. The main spike is now much narrower and the leakage from it has been reduced so that the component of the signal at 33 Hz is clearly visible.

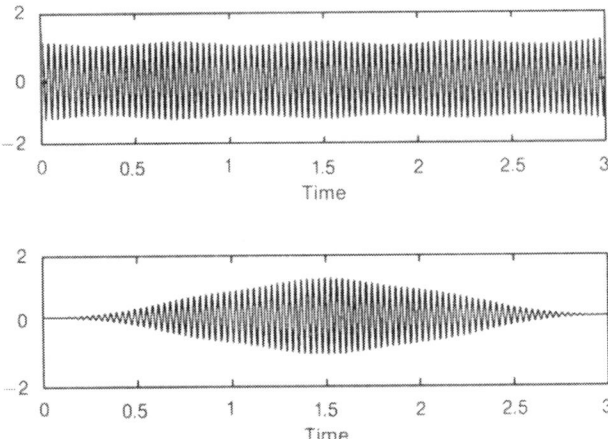

Fig. 11.18 Time-history before and after windowing.

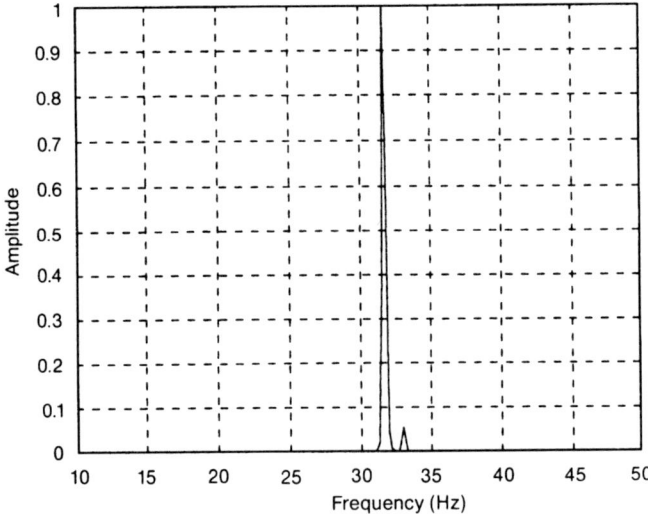

Fig. 11.19 Spectrum of windowed signal.

When using a window it is important to bear in mind that you are removing information from the signal that may be useful. For instance, many transients have the property that they are zero at the beginning and end of the time record. Recalling that we introduced windowing in the case of continuous signals to force the time record to be zero at the beginning and end of the data, we see that for this kind of signal there is no need to window the data.

Consider a transient signal, such as that shown in Fig. 11.20. If the Hanning window shown in Fig. 11.21 is used on the data, we get the highly

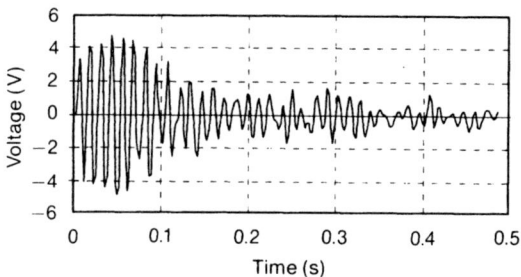

Fig. 11.20 Transient signal due to impacting a concrete beam.

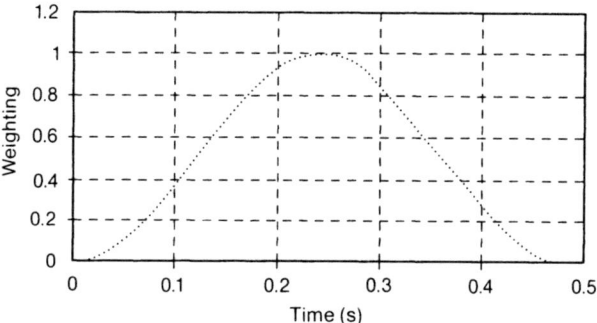

Fig. 11.21 The Hanning window used (incorrectly) to modify the transient signal.

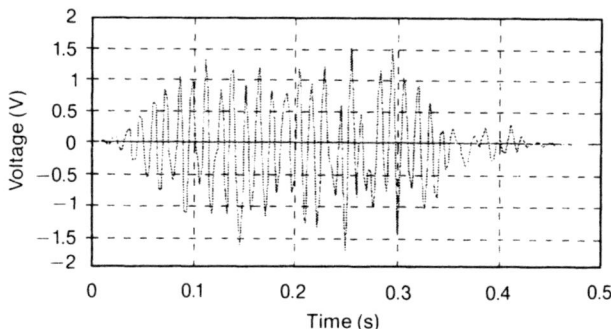

Fig. 11.22 The signal of Fig. 11.20 after windowing.

distorted signal shown in Fig. 11.22. In many transients much of the energy in higher modes occurs in the early part of the record and is likely to be lost after windowing. In general windowing transient signals is dangerous and should be avoided.

References

[1] Lynn, P.A. *An Introduction to the Analysis and Processing of Signals.* Macmillan, UK. 1982.
[2] Turner, J.D. and Pretlove, A.J. *Acoustics for Engineers* Macmillan, UK. 1991.

Index